Electron beam welding

Dipl.-Ing. H. Schultz

VERLAG DVS

ABINGTON PUBLISHING

Woodhead Publishing Ltd in association with The Welding Institute

Cambridge England

Published by Abington Publishing,
Woodhead Publishing Limited, Abington Hall, Abington,
Cambridge CB1 6AH, England
www.woodhead-publishing.com

First published in German, 1989, DVS, ISBN 3-87155-111-2
First published in English 1993, Woodhead Publishing Ltd
Reprinted 2000, 2002, 2003, 2004

English translation sponsored by:
Messer Griesheim GmbH
Steigerwald Strahltechnik
Benzstraße 11
D-82178 Puchheim
Germany
Tel (-89) 8092-0, Fax (089) 8092-215

British Library Cataloguing in Publication Data
A catalogue record for this book is available from the British Library

ISBN 1 85573 050 2

Printed in the United Kingdom by Lightning Source UK Ltd

Contents

Preface

The literature on electron beam welding is very extensive and increasing at the rate of about a hundred new publications a year. In spite of this, there is as yet no book which summarises all the processes and machines employed, and describes the many uses to which these are put.

This book attempts to fill this gap without aspiring to be a complete scientific description of the mechanisms and processes involved. I have set myself the task of explaining the trials, developments and manufacturing processes involved in electron beam welding in a practically orientated way aimed specifically at the average engineer. In doing so, the current state of the technology is described, not only with reference to the processes used but also including a great deal of basic welding information which should help the engineer in his own deliberations. Our highly developed and rapidly changing world demands that today's engineer has a broad general knowledge to enable him as a specialist to solve both present and future problems successfully.

I would like to express my thanks both to my numerous colleagues and to the companies which have assisted me, and especially to Messer Griesheim GmbH Steigerwald Strahltechnik/FRG, for their generous support and for their advice in the drafting of the manuscript. I also wish to thank them for sponsoring the translation of the book into English and thereby making it accessible to an international readership.

My special thanks go posthumously to Dr mont Walther Hiller whose valuable groundwork was an encouragement to me to write this book.

I would also like to thank the German Welding Society (DVS) and the Dokumentation Schweißtechnik for their valuable assistance and willingness to help in obtaining specialist literature.

Krailling, April 1993 *Helmut Schultz*

List of abbreviations

a	Shortest distance between pores or voids	mm
a	Thickness of fillet weld	mm
a_x; a_y; a_z	Deviation of the electron beam from the joint or of the focus from the surface of the position of welding in the X-, Y- and Z-directions as a measure of the accuracy of guidance	mm
Δa_1	Weld convexity	mm
Δa_2	Root convexity	mm
Δa_3	Depth of undercut	mm
A_A	Working distance	mm
A_F	Focal distance	mm
b	Width of weld	mm
b_A	Amplitude of beam oscillation	mm
b_P	Width of beam oscillation	mm
b_S	Width of weld gap	mm
b_Z	Width of fusion zone at any position	mm
d	Diameter	mm
d_{Fl}	Beam spot diameter	mm
d_{Fo}	Focal diameter of beam	mm
d_R	Diameter of circumferential weld	mm
d_{St}	Beam diameter at any position	mm
e	Edge misalignment	mm
f	Frequency of oscillation	Hz
F	Force	N
F	Greatest cross sectional area of pores or voids	mm^2
F_{Fl}	Area of the beam spot at the point of impingement	mm^2
F_{Fo}	Area of focal spot of beam	mm^2
g	Gravitational acceleration	$m \cdot s^{-2}$
G	Magnetic flux density	$V \cdot s \cdot m^{-2}$
h	Height of weld seam	mm
i_p	Pulse pause time	ms
i_f	Pulse frequency	Hz
i_t	Pulse duration	ms
i_v	Pulse ratio	–
I_A	Beam impingement current	mA
I_D	Transmitted beam current	mA
I_H	Cathode heating current	A
I_K	Beam deflection current	mA
I_L	Lens current	A
I_P	Beam oscillating current	mA
I_S	Beam current	mA
I_{SD}	Continuous beam current	mA
I_{SP}	Pulsed beam current	mA
j_e	Emission current density	$A \cdot cm^{-2}$
j_{eR}	Emission current density limited by space charge	$A \cdot cm^{-2}$

j_{eT}	Temperature dependent emission current density	$A \cdot cm^{-2}$
k	Distance between the cathode and the control electrode	mm
I	Length of weld seam	mm
L	Power density	$W \cdot cm^{-2}$
p_A	Working pressure, the pressure in the working chamber	mbar
P_E	Pressure in the beam generator	mbar
P	Beam power	kW
Q	Vacuum leakage rate	$mbar \cdot dm^3 \cdot s^{-1}$
R	Brightness	$A \cdot cm^{-2} \cdot sr^{-1}$
R_a	Centreline roughness	µm
R_K	Brightness at cathode	$A \cdot cm^{-2} \cdot sr^{-1}$
R_L	Radiance	$W \cdot cm^{-2} \cdot sr^{-1}$
$R_{p0.2}$	0.2% proof stress	$N \cdot mm^{-2}$
R_W	Brightness at workpiece	$A \cdot cm^{-2} \cdot sr^{-1}$
s	Depth of fusion zone	mm
t	Thickness of workpiece	mm
t	Strength determining thickness of the weld	mm
t	Time	s
T	Temperature	°C; K
u	Focal shift in the axis of the beam	mm
U_B	Accelerating voltage	kV
U_{St}	Control voltage	kV
v	Beam offset	mm
v_s	Welding speed	$mm \cdot s^{-1}$; $m \cdot min^{-1}$
V	Volume of working chamber	m^3
X,Y,Z	System of co-ordinates	–
z	Crossover shifting	mm
z	Thickness of the centering lip	mm
α	Aperture angle	rd
α	Tangential angle to weld seam	°
α_K	Aperture angle towards cathode	rd
α_W	Aperture angle towards workpiece	rd
β	Angle of deflection	°
γ	Density	$g \cdot cm^{-3}$
δ	Gradient of the integral curve	°
σ	Standard deviation	-
ω	Solid angle	sr

1 Introduction

1.1 History

Nowadays, the technical achievements which make life so convenient are taken for granted, and the invention, development and testing carried out by our predecessors are often overlooked. Although an electron beam welding machine is of course not an item of every day use, it can be used to manufacture products which in one way or another add to the high standard of living experienced today. Over a hundred years have passed since the first experiments by the physicists Hittorf and Crookes to produce cathode rays in gases (1869) and to use them to melt metals (1879) which have led to today's CNC manufacturing machines for electron beam welding of jet engine components, to name just one example of many.

Initially cathode rays were little more than an interesting physical phenomenon which, in 1895, led to the discovery of one particular type of radiation by Röntgen and which was described by Thompson (1897) and Millikan (1905) as 'rapidly moved electrons'. As far as processing of material was concerned, such beams were of no significance. Indeed quite the opposite was the case since, in all the experiments carried out at that time, the heat produced by the collision of the electrons with the anode was regarded as a great disadvantage and continued attempts were made using water cooling to prevent the anode target from being melted [1]. It was Marcello von Pirani[1)] who first thought of making use of this effect by adapting a cathode ray to construct a type of electron beam furnace (Fig. 1) for melting tantalum powder and other metals, and which was patented in 1905 and 1907.

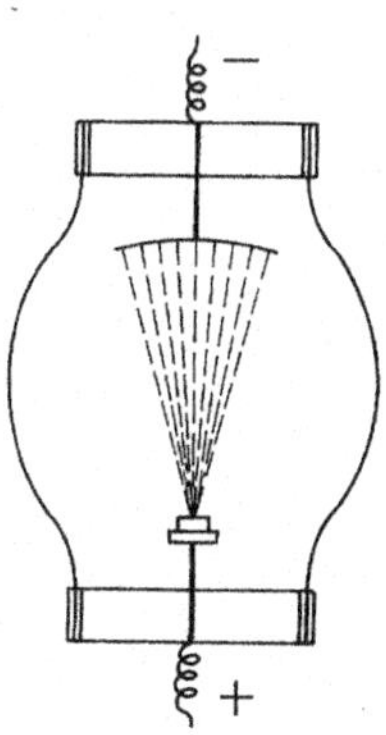

Fig. 1 Extract from a patent by Marcello von Pirani regarding the 'Production of homogeneous bodies from tantalum or other metals' taken out on the 26 March 1907 [1].

[1)] Pirani, M. V., 1880–1968, German physicist

In the following decades many scientists became involved with electron beams. Amongst others, Langmuir, Child, Richardson, Dushman and Wehnelt investigated the laws governing the production of such beams, whilst Busch, Rogowski, Flegler, Davisson, Calbrick and others worked on the basis of electron optics. The first important uses of electron beams were in the construction of oscilloscopes and microscopes. Von Ardenne (1938) used them for drilling metals, and together with Rühle (1939) for melting and vaporising metals. Sufficiently powerful vacuum pumps were still unavailable, however, to permit use of such beams in larger scale industrial applications.

A new epoch in the working of materials using electron beams began in 1948. It occurred to the physicist Steigerwald[2)], who at this time was involved in development of more powerful beam sources for use in electron microscopes, that electron beams could be employed as thermic tools, in particular for drilling precious stone bearings for watches and clocks, for drilling wire drawing dies and for soldering, melting and welding metals under vacuum [2].

The results of the first trials proved very promising and led to the signing of a licensing agreement with an interested American party. At that time electron beam welding was thought of in the same terms as arc or gas welding, namely as a process using a suitable source of heat, which had the particular advantage that gas sensitive metals were protected from reaction with the atmosphere. The breakthrough in industrial electron beam welding came in 1958 with the requirement to butt weld together 5 mm thick plates of Zircaloy [3]. By gradually increasing the beam current it finally became possible to completely penetrate the full thickness of the workpiece to produce weld seams which, as was required, were of considerably greater depth than they were wide. This deep penetration welding process immediately drew worldwide interest, but it was in the USA that its real technical significance was most rapidly perceived.

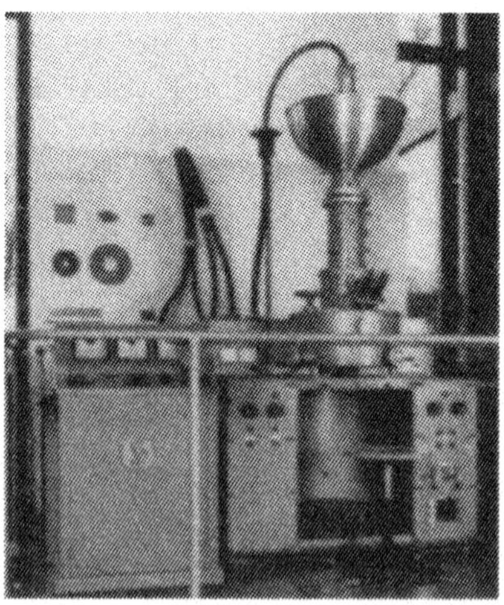

Fig. 2 The first electron beam welding machine for deep welding (1958), $U_B = 55$ kV, $I_S = 20$ mA, at present in the Deutsches Museum, Munich.

[2)] Steigerwald, K. H., 1920, German physicist

Following this success Steigerwald sold two such electron beam machines. One was installed in Pittsburg/USA for welding submarine components, whilst the other was used for many years in industry in Germany, and today can be seen in the Deutsche Museum in Munich (Fig. 2).

After the discovery of the deep welding effect an upsurge in development of new machines occurred, in particular in France and Great Britain [4]. Initially the electron beam had been considered only in relation to the surface of the material, but now attempts were made to increase the power density and beam current in order to be able to weld even thicker components. The nuclear and aerospace industries were the first to control the use of electron beam welding. Amongst the milestones in the subsequent development of the process and equipment were coupling of the high voltages in the beam chamber without the use of insulating oil, changing the cathode using a clamping system, isolating the vacuum in the beam gun from that in the working chamber, the construction of welding machines with much higher beam powers and larger working chambers, and the use of cycle type welding machines for mass production techniques. These all led to a multitude of new applications for electron beam welding. Today, even specialists have difficulty in appreciating the wide field of use of electron beam welding in materials processing.

1.2 Special characteristics of electron beam welding

As an introduction, a number of the special characteristics of electron beam welding are summarised and discussed in greater detail in the following sections. In comparison with other methods of joining, electron beam welding is characterised by:

- an extremely high power density of about $10^7\,W \cdot cm^{-2}$ at the focus of the beam, Table 1;
- energy transfer occurs not by conduction of heat across the surface of the workpiece, but much more efficiently within the workpiece itself;
- as no edge preparation is necessary, regardless of the thickness of workpiece, filler metal is not required;
- the high welding speed results in narrow welds and heat affected zones with little distortion of the workpiece;
- inertia free oscillation of the electron beam makes it possible in many cases to join materials otherwise considered unsuitable for welding;
- the variable working distance allows workpieces of widely differing shapes to be welded;
- as welding is carried out under vacuum, no consumables (gases, fluxes) are required to protect the weld pool from oxidation;
- short evacuating times can be achieved by adapting the working chamber to suit the number and size of the workpieces;
- computer monitoring and control of the electrical and mechanical welding parameters is possible;
- the welding parameters, and thus the quality of the welds produced, are highly reproducible and consistent;

- at accelerating voltages above 60 kV lead shielding is directly bonded to the welding machine in order to prevent X-ray emission.

Table 1 The maximum power densities of the various heat sources used for welding

Heat source	Maximum power density $W \cdot cm^{-2}$
Gas flame	$5 \cdot 10^3$
Electric arc	10^4
Plasma	10^5
Laser beam (continuous)	10^7
Electron beam	10^7

Unlike other manufacturing processes, a relatively limited range of electron beam welding machines is capable of meeting the various power requirements of the most diverse applications:

- beam powers of far less than 1 kW up to 300 kW are available for welding material thicknesses of less than 0.5 mm up to 300 mm [5];
- machines are available for welding variable one-off components as well as for use in mass production operations as in the automotive industry;
- simple longitudinal weld seams can be welded as well as complicated three dimensional components requiring use of programmed welding parameters and workpiece manipulation;
- structural steels, alloy steels, non-ferrous metals and even gas sensitive special metals can be successfully welded.

Unfortunately there are no statistics available showing exactly how many electron beam welding machines are in use world-wide. In 1979 it was estimated that 3000 machines were in use [6]. It can be said with certainty, however, that the economic viability of every electron beam welding machine sold today is proven, and that it is this single fact which is the best possible argument for the use of this process.

1.3 Other beam welding processes

Laser beams, best known for their applications in measurement, eye surgery and for data transfer, etc, have now also found a place in cutting and welding technology. A laser produces a monochromatic (one wavelength) and coherent (in phase) beam of light with the aid of particular laser active media (gases, liquids, solids). The medium is excited to a higher energy state such that an incident quantum of light will cause emission. Mirrors are then used both to arrange and amplify the stimulated emission into a collimated beam. One of the properties of a laser beam resulting from this method of producing light is the extremely low divergence of the beam. The laser beam can thus be transmitted across large distances and can be focused to very small diameters with high power densities. Thus, using the deep welding effect, laser beams are also capable of producing weld seams which are considerable deeper than they are wide. However, how in detail the energy of the laser beam is transferred into the material is not yet fully understood. An electron transfers its kinetic energy to the material over a distance of about 0.01 mm, largely in the form of heat, and the same process appears to occur in the case of laser beams, but within a distance of as little as 0.01 µm. In addition, processes such as the absorption occurring in the ionised metal vapour (plasma) also play a significant role in transferring energy.

For welding applications, CO_2 is the main laser active medium. In contrast to solid media, it is particularly suited to producing high continuous beam powers. Laser welding is normally carried out in air. Therefore there is no need for a working chamber which has to be evacuated. But it is necessary, as with other welding processes, to protect the weld pool from ingress of the surrounding atmosphere. In laser welding, both argon and helium separately or a mixture of both are used as the shielding gas. Even without an evacuated working chamber, the capital cost of high power laser welding machines is considerable, so that most laser welding machines at present in use in the automotive and electro-technical industry have a beam power of less than 5 kW.

Prior to the introduction of laser beam welding, light beam soldering and welding was also of some importance. The light was produced by a gas discharge lamp, collimated and focused on to the workpiece with the aid of elliptical mirrors. The relatively poor ability to focus such light, even with beam powers of 5 kW, permitted only thin components of ca 1 mm thickness to be soldered or welded [7].

Other trials, for example to weld with ion beams, have never progressed beyond the laboratory stage and today are no longer of any real practical significance [8].

2 Generation of the electron beam

2.1 Free electrons

If one considers the physics of the electron beam process it may seem surprising that the heat for this type of welding is provided by the kinetic energy of particles which are among the lightest present in matter. The fact that the numeric value of the electron rest mass is only $9.1 \cdot 10^{-28}$ g or that an electron is about 1836 times lighter than an identically charged proton particle is, as most atomic dimensions, beyond our everyday comprehension. The apparent disadvantage of using a particle of low mass for a kinetic energy transformation process is made up for by one of the electron's most important properties: it carries an electrical charge. Its magnitude, $1.6 \cdot 10^{-19}$ As, is just as incomprehensible and is mentioned here only for completeness because this charge ultimately results in the beam current. Much more important is the fact that this charge makes it possible to accelerate electrons by electrical fields so

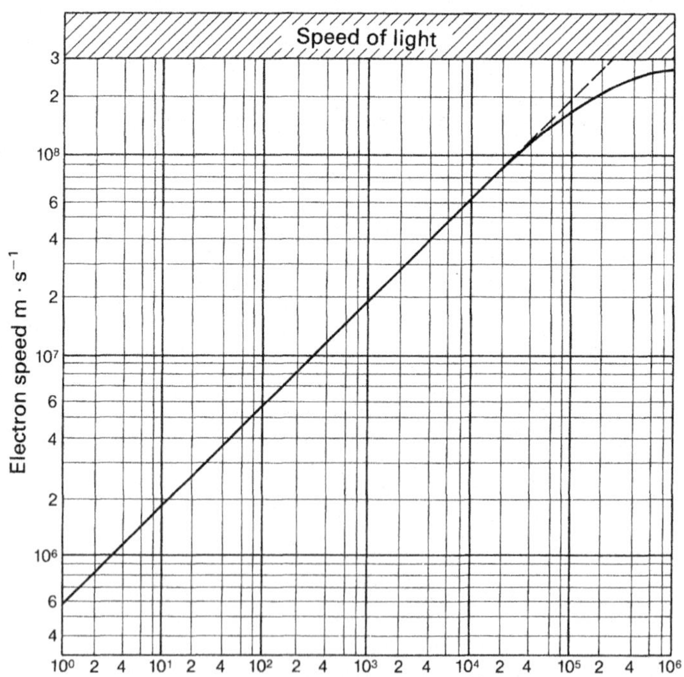

Fig. 3 Relationship between the electron speed and the accelerating voltage.

effectively that they ultimately gain the amount of energy required to weld metals. As can be seen from Fig. 3, at a typical accelerating voltage used for welding of $150\,kV = 1.5 \cdot 10^5\,V$, and in a suitable vacuum, electrons reach a speed of $2 \cdot 10^8\,m \cdot s^{-1}$, or in other words two thirds the speed of light. The kinetic energy of the electrons, which is given by the term $\frac{1}{2}mv^2$ is supplemented by the relative mass increase of the electrons of about 35% which becomes noticeable at this velocity [9].

It is well known that electrons are found in the shells surrounding atoms in which they occupy discrete energy levels. In the case of metals, however, the outer electrons are bound only weakly to the core of the atom. Such electrons are more or less free to move throughout the crystal lattice, and this gives metals their excellent electrical conductivity (conducting electrons). These electrons are unable to escape from the surface of the metal, however, since they are bound to the crystal lattice by a potential barrier which can only be overcome if additional energy, the so-called work function, is provided. The most technically reliable way of supplying this energy is to raise the temperature of the metal. This increases the energy of the free electrons, so that they can overcome the potential barrier and escape from the surface of the metal. By this process of thermal emission a metal becomes a source of electrons and, as explained later, by applying a negative voltage, it can provide the source of an electron beam.

2.2 Cathode

Far from being just a source of electrons, the cathode also performs a number of other important functions. It is required to emit a large number of electrons for a long time while consuming as little thermal energy as possible; by using little heating power; in other words it should be able to provide high beam currents.

According to Richardsons rule[3)], to obtain a high beam current the emitting surface of the cathode should be at a high temperature. Because of the exponential relationship between the cathode temperature and the electron emission, expressed as a maximum achievable emission current density or temperature determined saturation current density j_{eT} in $A \cdot cm^{-2}$, it is particularly effective to choose cathode materials with high melting points. For this purpose, in addition to tantalum, tungsten partially alloyed with rhenium is also particularly suitable because of its comparatively low vapour pressure under vacuum which ensures a relatively long life time. In electron beam welding machines, emission current densities of about $j_{eT} = 5\,A \cdot cm^{-2}$ are commonly achieved with this type of directly heated tungsten cathode.

The heating power to raise the cathode to the required temperature depends not only on its dimensions but also on its shape and the type of heating used. Directly heated strip cathodes just mentioned (Fig. 4) have proven themselves for use in electron beam welding machines for many reasons. This type of

[3)] Richardson, O. W. 1879–1959, English physicist

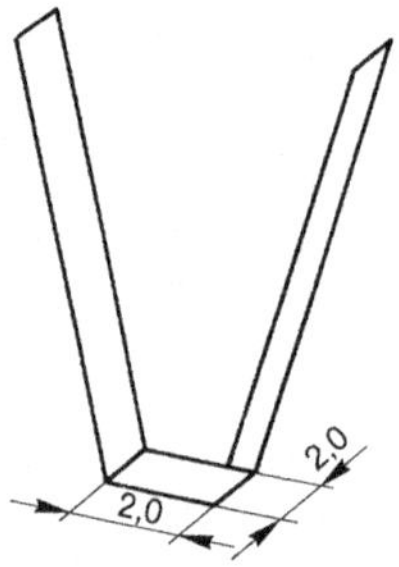

Fig. 4 Strip cathode for beam powers of up to P = 15 kW.

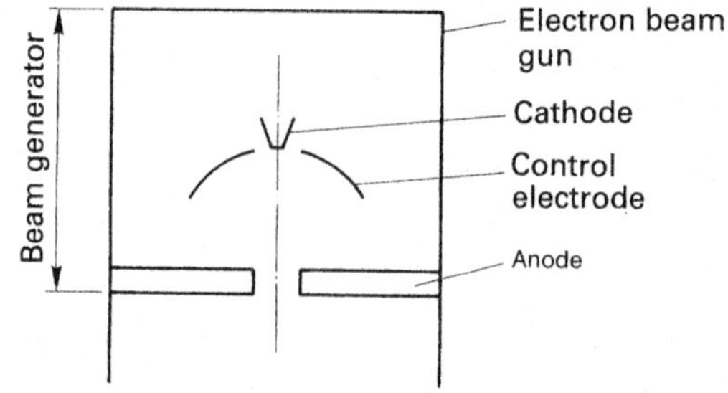

Fig. 5 Diagram of a triode beam generator system.

cathode is connected as a resistor in the electrical circuit and is heated directly by the passing current. Strip cathodes have a high heating efficiency and, in conjunction with a particular type of cathode holder, provide a very geometrically stable emission surface and thus very stable beam parameters.

As the only part of the electron gun which wears, cathodes are easily replaced, as explained in section 8.2. Because of their low mass, directly heated cathodes cool rapidly once the heating current is switched off. This is itself beneficial in preventing unnecessary oxidation of the cathode if the electron beam gun is vented prematurely.

Indirectly heated cathodes are heated by electron impingement with the aid of an additional element (auxiliary cathode). As they do not carry a heating current they can be more massive and have a more complex shape. The best known of this type are bolt cathodes which are used to a limited extent in certain types of low accelerating voltage electron beam welding machines. Bolt cathodes are easier to manufacture and have a longer life than strip cathodes. However, even after only a few hours of welding the emitting surface of the cathode sustains considerable damage in the form of craters caused by ion impingement. This results in a change of the beam characteristics and hence in the welding results. These changes have to be compensated for by suitable adjustments of the control settings over the life time of the cathode. A further disadvantage is relatively rapid deterioration of the auxiliary cathode, which must be replaced much more frequently than the bolt cathode.

2.3 Anode

By thermal emission the cathode provides a supply of free electrons, but these do not form an electron beam which can be used for welding. Because of the thermal emission, the cathode itself becomes positively charged, which attracts the negatively charged electrons causing them to collect around the cathode in the form of a cloud. To give the electrons the kinetic energy they require, they must be accelerated to a high speed, and to achieve this a negative voltage is applied to the cathode and an annular anode at earth potential is mounted above the electron cloud (Fig. 5). The electrical field between anode and cathode accelerates the electrons in the direction of the anode. They pass through the anode with a constant velocity. At the same time a high voltage generator supplies the cathode continuously with current, or more simply with new electrons. Thus an electron current flows in a closed circuit from the workpiece, through the machine and the earth connection back to the high voltage generator.

2.4 Control electrode

With a cathode and an anode the simplest form of beam generator can be constructed. This 'diode' system, also called a Pierce system[4)], was in fact used in a number of machines in the initial years of electron beam welding.

It does have one considerable disadvantage, however, in that the beam current can be controlled only by altering the accelerating voltage or the temperature of the cathode. This dependency is quite unsuitable for economic welding. After a relatively short period of use of this type of system, only 'triode' systems consisting of an anode, cathode and control electrode, also known as a Wehnelt Cylinder[5)], were used in welding and other electron beam manufacturing machines. With the aid of a control electrode the electron current and thus the beam current can be adjusted as follows. The control electrode is at cathode potential but has a higher negative voltage superimposed on it from a separate power supply. In accordance with the physical law that like charges repel, the control electrode is, at a suitable potential, capable of repelling the electrons back towards the cathode in spite of the potential difference between the cathode and the anode, thus shutting off the beam current by reducing the effective emitting surface to zero. If the control voltage is reduced, the electron emitting surface will grow and the beam increases (Fig. 6*b* and 6*c*). Adjustment of the beam current is thus achieved by varying the control voltage: high control voltage = low beam current, low control voltage = high beam current. Figure 6*c* illustrates the greatest increase in the emitting surface which can be achieved.

By further decreasing the control voltage U_{St}, the emitting surface expands to the edges of the face of the cathode, and eventually wraps over the edges of the cathode (Fig. 6*d*) and excessive divergence or distortion of the beam occurs. Thus, to produce a rotationally symmetrical electron beam there is a minimum permissible value, dependent upon the accelerating voltage and the triode arrangement, below which the control voltage must not fall.

Figure 7 shows the shape of the electron beam in a triode system as typically used in electron beam welding machines. Of note is the typical necking of the electron beam just beneath the cathode's emitting surface which results both from the geometry as well as the voltage applied to the control electrode. This necking is known as the position of real crossover. Once the electron beam passes through the anode and into a region free of electrical fields, its

[4)] Pierce, G. W., 1872–1956, German physicist
[5)] Wehnelt, A. R. 1871–1944, German physicist

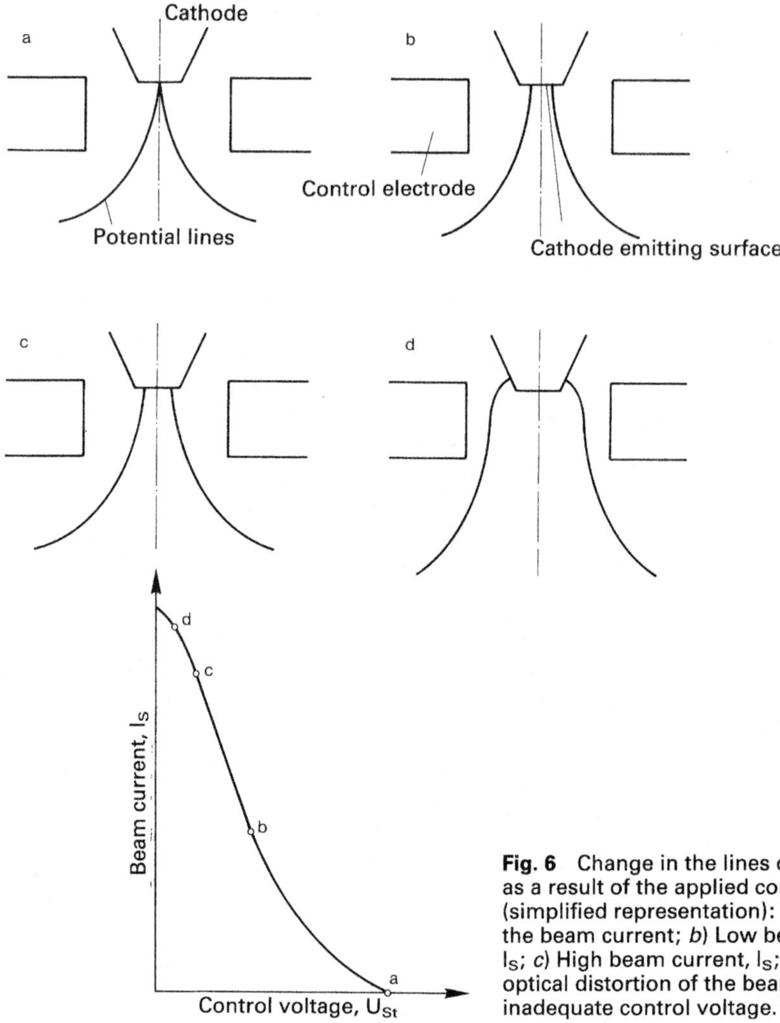

Fig. 6 Change in the lines of potential as a result of the applied control voltage (simplified representation): *a)* Blocking the beam current; *b)* Low beam current, I_S; *c)* High beam current, I_S; *d)* Electron optical distortion of the beam because of inadequate control voltage.

geometric shape will not change until it reaches the focusing lens. Extending the tangents to the beam envelope backwards beyond the cathode leads to a point of intersection with the beam axis known as the point of virtual crossover, which is of particular significance in electron optical calculations. Figure 8 is an enlarged diagram showing somewhat more accurately the paths of the electrons under the effect of field distribution.

It can be seen that not only is the real crossover not in reality a single sharply defined point of intersection, but that it is made up of a series of differently intersecting lines.

As in light optics, in electron optics one also speaks of the focal length of a system. In particular, during the start of the development of this process one

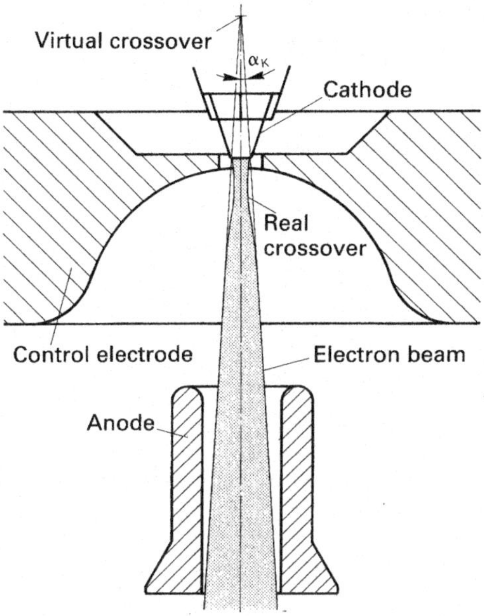

Fig. 7 Beam geometry in a triode system.

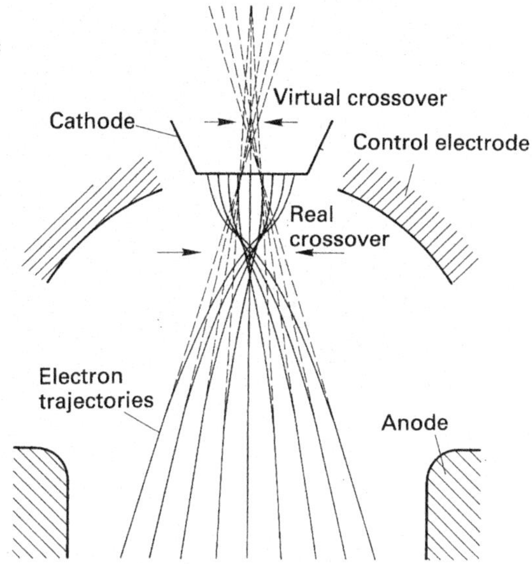

Fig. 8 Electron trajectories which can lead to occurrence of a real and virtual crossover.

differentiated between short focal length (Rogowski[6]) and long focal length (Steigerwald distant focus) systems. Today, electron beam welding machines are in the main equipped with modified Rogowski systems.

2.5 Spacial charging operation

The physical laws according to which the electrons move between the cathode, the control electrode and the anode are very complex but do not need to be explained individually in detail in order to gain an understanding of the further development of the systems. Only one particular relationship need be understood, this being the setting of the cathode heating current which is of particular practical importance.

When operating, the cathode is subject to a small yet significant loss of material by vaporisation and pulverisation, and thus has only a limited service life.

Vaporisation is a result of the high cathode temperature in a vacuum, even though the temperature is still significantly below that of the melting point of the cathode material. Pulverisation is caused by the impact of high energy ions (Fig. 9) which are produced by the electron beam on impact with residual gas molecules as it passes through the vacuum and with vaporised metal at the point of welding. This slow progressive loss of material reduces the cross

[6] Rogowski, W., 1881–1947, German physicist

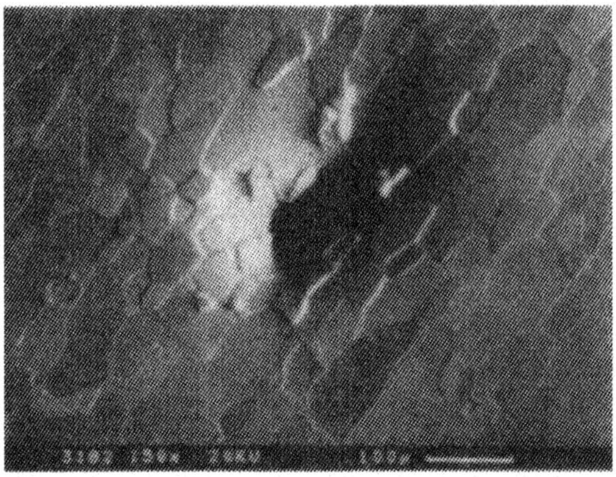

Fig. 9 Emission surface of a strip cathode torn open by ion bombardment.

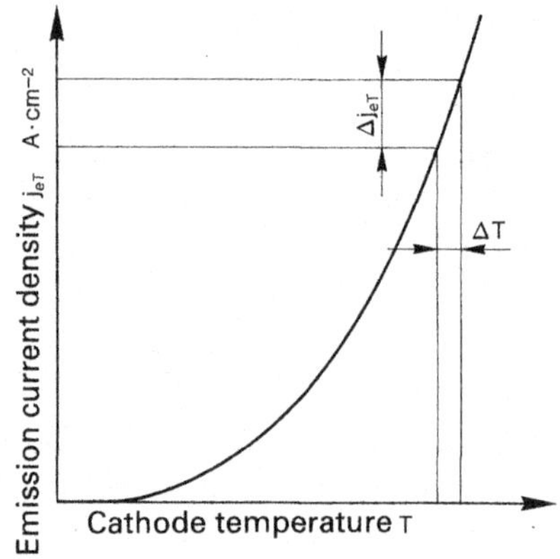

Fig. 10 Large changes in the emission density j_{eT} resulting from small changes in the cathode temperature T.

section of the emitting area of the cathode, the heating current reduces and thus simultaneously the temperature of the cathode falls. According to the Richardson relationship already mentioned, the emission current density j_{eT} is strongly dependent upon the temperature of the cathode such that low variations in temperature during the life of the cathode can have a considerable effect on the beam current (Fig. 10). For electron beam welding with stringent requirements in terms of weld seam reproducibility, this type of dependency would be extremely detrimental. In such a case it is possible cleverly to use another physical relationship to compensate for this, namely the Langmuir relationship[7)]. According to this, a given accelerating voltage, as a function of a number of geometrical characteristics such as the distance of the cathode from the anode, etc, is only able to extract a given maximum electron current, the space charge limited emission current, j_{eR}, from the cathode.

Thus a sufficiently high cathode temperature must be set by adjusting the heating current so that even at the maximum accelerating voltage sufficient electrons are still available. In other words, j_{eT} must always be larger than j_{eR}. Under these conditions the cathode will be surrounded by a cloud of these 'superfluous' electrons which has an inherent charge and which thus limits any further emission of electrons. The cathode works in space charging mode. Figure 11 shows schematically how, at an adequate cathode temperature, the beam current remains unaffected by variations in the temperature dependent emission current. In practice a series of monitor and control units fitted to the welding machine makes it easy to monitor and set the spacial charging region, and to ensure that the life of the cathode is not reduced significantly. This is explained further in section 7.4. For completeness it must also be mentioned that the life of the cathode is also determined by other factors, and in particular by the beam current, the material to be welded and the pressure in the

[7)] Langmuir, J., 1881–1957, American chemist and physicist

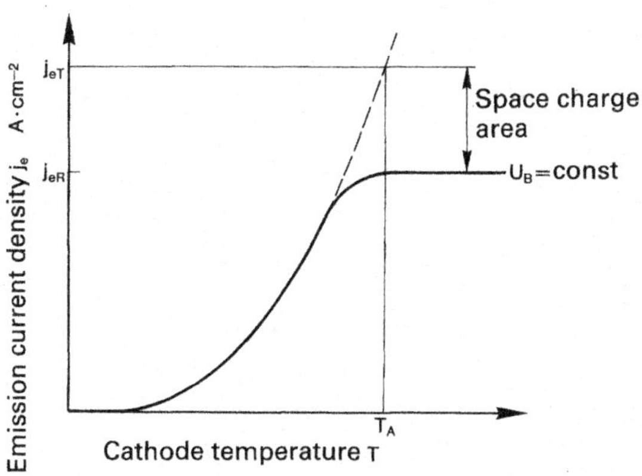

Fig. 11 Definition of the space charge area.

electron beam gun, etc. The average life of a cathode in a powerful electron beam machine can be assumed to be around 20 hours of welding.

2.6 Focusing lens

The geometric shape of the cathode, control electrode and anode not only have an effect on the more or less curved path of the electrons when in the electron gun, but also on the further path of the electrons to the point of welding.

As has already been mentioned, the electric field in the triode system shapes the beam to an initial focus, the crossover point, from which the electrons move through the anode, as like charge carriers repelling one another (Fig. 7). Although the electrons have been accelerated to their final velocity after passing through the anode, the divergent beam they make up still does not have sufficient power density for welding metals. To achieve this the electron beam must be focused.

Using an annular coil it is possible to influence the direction of movement of the negatively charged electrons by application of magnetic fields. The ring consists of a large number of windings of wire shielded on three sides by a high permeability iron casing (Fig. 12). A direct current flows through the windings of the annular coil producing a magnetic field which acts inwards from the iron casing on the electron beam, focusing it in the same way as a condenser lens is used to focus a beam of light. In doing so, it is irrelevant whether the direct current flows through the coil as shown in Fig. 12, or in the opposite direction. The electrons leave the magnetic lens, without experiencing any change in speed, describing a spiral slightly curved path, and meet at a focal point. Diameters of 0.1–1.0 mm are typical of electron beam welding, depending upon the beam power and the focal distance.

This small focal diameter achieves the power density of about $10^7\,W \cdot cm^{-2}$ required to weld metals of several centimetres thickness.

Without going into details which are explained in Chapter 4, for welding, the beam must be focused approximately on the surface of the workpiece. This is achieved by varying the current flowing through the annular coil. This current is the main factor which determines the power of refraction, or focal length of

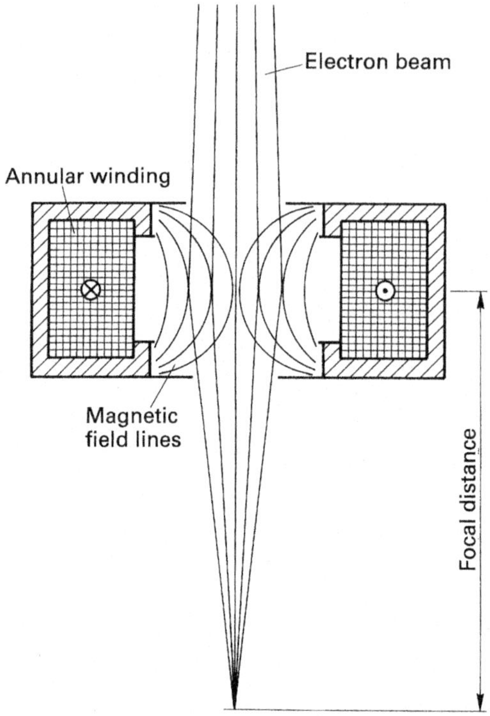

Fig. 12 An electromagnetic lens for focusing the electron beam.

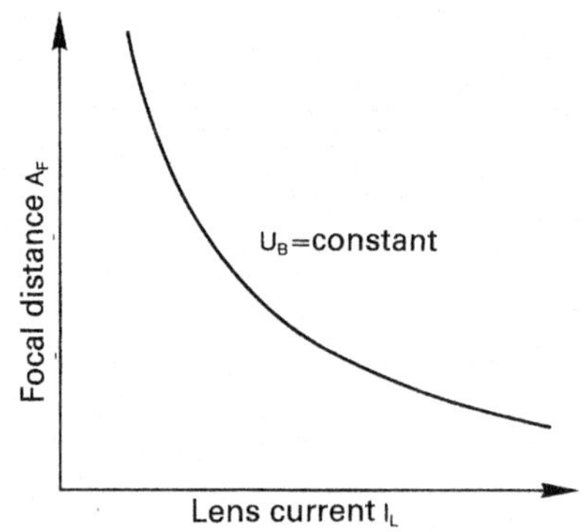

Fig. 13 Dependence of focal distance A_F on the lens current I_L at a constant accelerating voltage U_B.

the magnetic lens, and is referred to as the lens current, I_L. The focal length is inversely proportional to the square of the lens current, and is also directly proportional to the accelerating voltage (Fig. 13). In other words, the focal length may also be adjusted by altering the accelerating voltage. This however is not generally done in practice.

Graphically the electron beam impinges on the surface of the workpiece in the shape of a cone standing on its tip. The half angle enclosed by the tip of the cone is referred to as the angle of aperture α which plays an important role in the optical relationships of the system. In electron beam working processes one thus refers to an angle of aperture on the side of the workpiece of α_W, in this case the point of welding (Fig. 14), and an angle of aperture on the side of the cathode α_K at the point of crossover (Fig. 7).

In actual welding practice it is the aperture on the workpiece side, which can be altered within certain limits either by the focal distance or the lens current as well as by certain other parameters, which is of greatest importance. This angle, which on average measures about 10^{-2} rad or ca 0.6°, must be taken into account when welding with the beam close to a workpiece surface (Fig. 129, section 6.4).

So far it has been implied that the exact path of the electrons from the

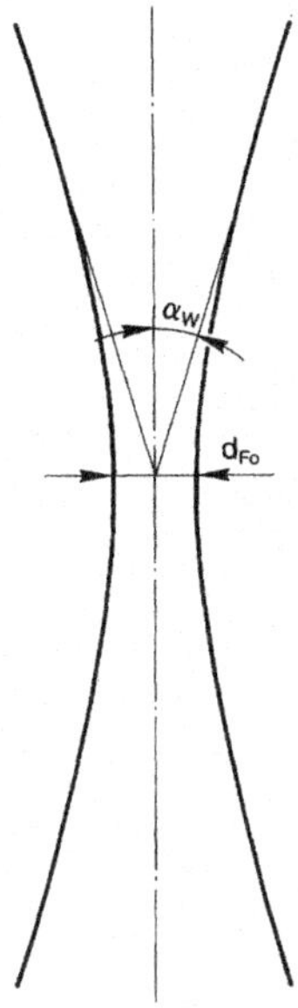

Fig. 14 Waisting of the beam (caustic) at focus showing the aperture angle towards the workpiece α_W and the focus diameter d_{Fo}.

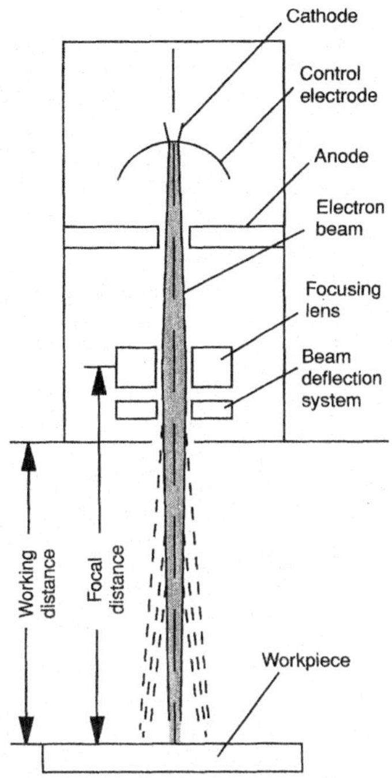

Fig. 15 An electron beam gun with focusing lens and beam deflection system.

cathode to the workpiece is actually more complex than has been described. Thus here also, as in light optics, a number of different types of imaging errors can occur which will influence the position and shape of the focal spot. The causes of these errors lie to some extent in the physical rules governing the generation and divergence of the electron beam (thermal divergence, space charge divergence, spherical aberration, astigmatism, scattering effects, etc). Further causes can be electrical and magnetic interference fields within the electron beam gun and the focusing lenses such that, instead of a sharp focus, only a waisting of the beam, called a caustic, is produced. This describes the envelope surrounding the paths of the outermost electrons in the beam (Fig. 14). As a rule, in electron beam welding machines, imaging errors are not corrected.

To correct such errors would require extensive extra instrumentation, whilst in reality such errors have such a minor effect on the welding process as a whole that correction is unnecessary [10]. One exception to this is correction of astigmatic errors, see section 2.8.

2.7 Deflection system

In most cases during welding the workpiece is moved along beneath the stationary electron beam. In other cases the electron beam gun itself moves,

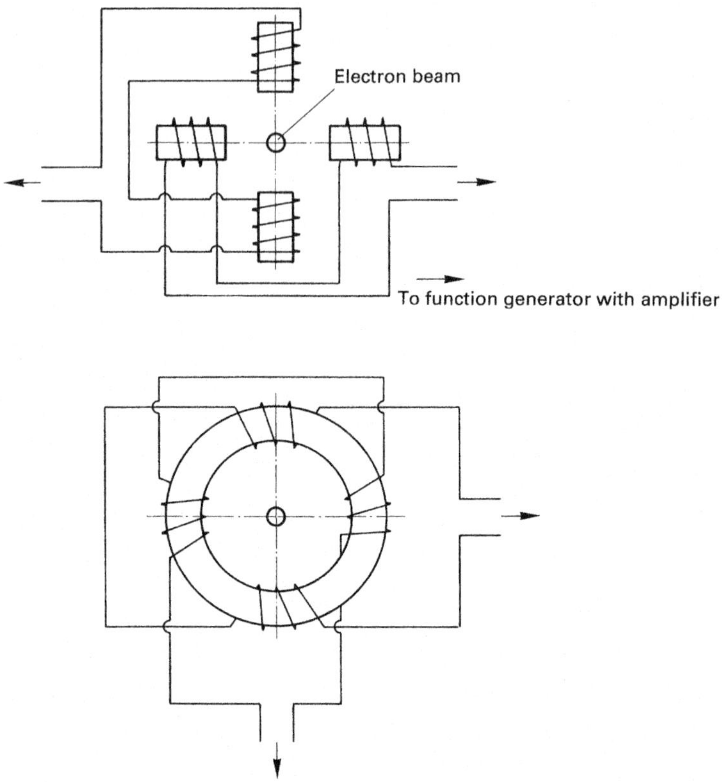

Fig. 16 A beam deflection system; left: main field coils; right: subsidiary field coils.

traversing the electron beam over the stationary workpiece. It is often necessary, however, to be able to move the electron beam itself from its normal undeflected axis, for example when welding T butts or to dynamically influence the welding process. To achieve this a deflection system is built into the electron gun beneath the focusing lenses (Fig. 15) which can deflect the beam by the application of a magnetic field. The deflection system consists of four coils wound on either a single, or number of, cores (Fig. 16), which are connected to a function generator and amplifier. The coils are paired across the main axes of the workpiece or movement of the electron gun such that by applying a suitable direct current the beam can be statically moved in either the X- or Y-directions or through any required angle in between.

Using an alternating current, the beam can also be made to oscillate. By controlling the frequency and phase difference of the current supplied to both pairs of coils it is possible to oscillate the beam through any desired path. This is discussed further in section 4.6. Thus, as the electron beam itself is inertia free and fully controllable, there is no limit to the frequency and shape of oscillation. The deflection system, however, can only impose high frequencies on to the electron beam without causing distortion if its own inductance is low.

The deflection systems of modern electron beam welding machines fulfill these requirements by using particular materials for the cores and by employing special methods of winding.

2.8 Beam correction systems

Great demands are placed on the focusing and positional stability of the focal spot. Thus in all positions the beam spot should be almost circular and its position should not alter, that is, the electron beam should remain free of imaging errors normal to the surface of the workpiece (Fig. 17). Furthermore, it is expected that the position of the spot should remain constant over the whole life of the cathode. Electron beam welding machines are therefore equipped with a number of additional pieces of instrumentation which are used to correct any deviations occurring in the electron beam.

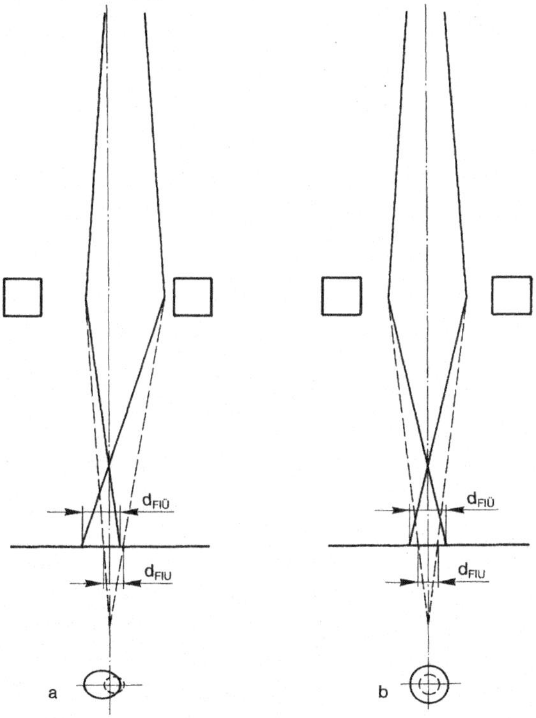

Fig. 17 Change in the diameter of the spot when over- ($d_{FIÜ}$) and under (d_{FLU}) focused: *a*) Beam axis off centre and inclined; *b*) Beam axis centred and at normal.

In the electron beam gun various precautions are taken to ensure that the electron beam, the control electrode and the anode remain centrally aligned even after a cathode change. Further correction of the beam is possible using the stigmator, which fulfills the function of compensating for the effects of electrical and magnetic interference, the astigmatism of the beam. The typical characteristic of astigmatism is an elliptical, oblique shape of the beam spot when altering the axial direction on changing the position of focus (Fig. 18).

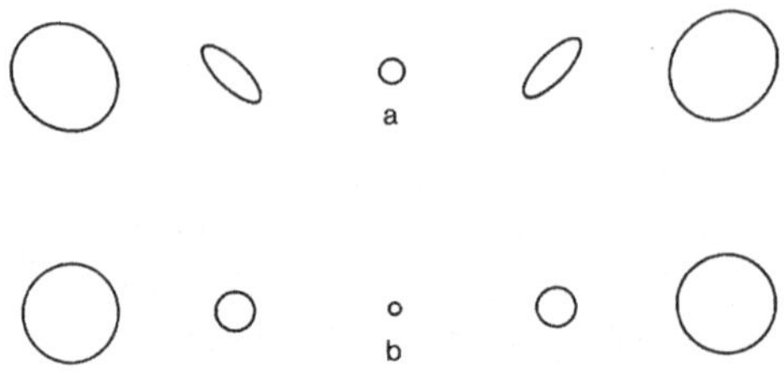

Fig. 18 Changes in the shape of the spot when over and under focused, for an electron beam: *a)* With astigmatic aberration; *b)* Without aberration.

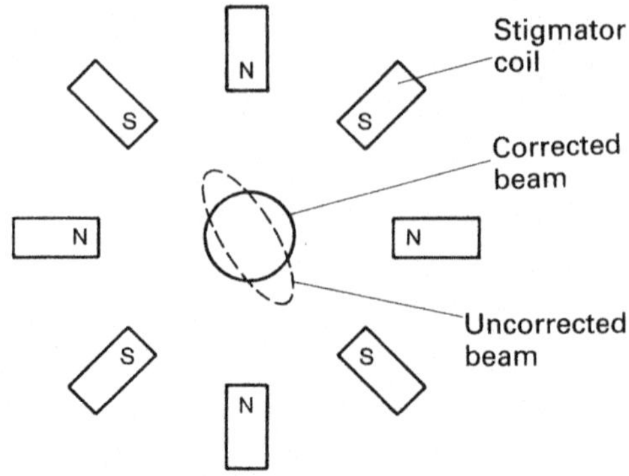

Fig. 19 A stigmator and the correction of an electron beam showing astigmatic aberration.

The stigmator is similar in construction to the deflection system, but differs from it in having other annular windings, polarities and to some extent in the number of poles (Fig. 19). Four poles can be controlled individually from a single power supply. Correction of astigmatic aberrations requires a particular approach, explained in the machine manufacturer's operating instructions.

2.9 Brightness

At the inception of electron beam welding technology, the opinion was that the accelerating voltage U_B, beam current I_S, welding speed v_S and beam focus diameter d_{Fo} parameters were all that were required to determine the width and depth of the melting zone produced in a given material. However, more detailed experimental investigations soon revealed that the relationships were in fact far more complicated than at first thought.

Even taking into account various different energy transfer efficiencies from the beam to the material being melted, the calculated results often differed from those actually measured in practice. It was realised that not only were the welding parameters which could be set on the operating panel of the machine of importance, but that also the shape of the electron beam itself would have to be taken into account. This led to use of the term 'brightness' from light and electron optics in electron beam welding [11–13]. Relative to the actual welding of a workpiece, the mean brightness R_W can be defined simply as the beam current I_S which impinges on an area F at a solid angle ω (Fig. 20), or more accurately, the quotient of I_S and the product $\omega \cdot F$. Occasionally, the expression 'radiance' R_L is also used, in which in place of the beam current I_S, the beam power P is included [12]. The magnitude of the beam power value of an electron optical system is determined by the electron gun. This value R_K can be increased by altering the emission current density j_{eR} if the electrons required to form the beam are drawn from the smallest possible area of

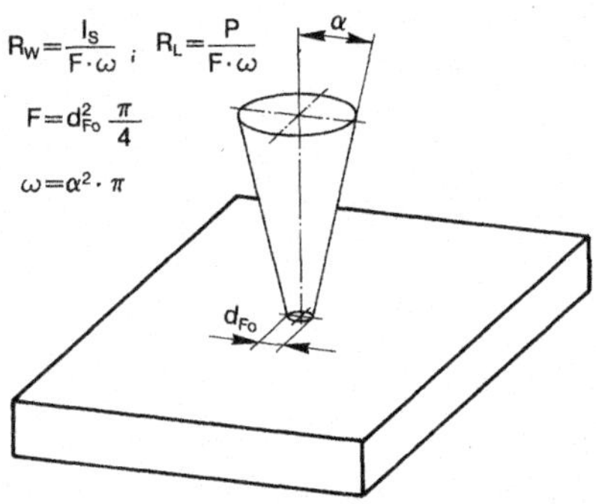

Fig. 20 Definition of the brightness of the beam R_W and of the radiance R_L.

emission on the cathode [14–16]. Further along the beam path the brightness R_K can at best be maintained, any increase no longer being possible. In practice R_W is generally smaller than R_K.

With the introduction of the geometrical solid angle parameter ω it can now be shown that a proportional relationship between the brightness R_W and the weld profile exists, expressed by the ratio of the depth of the fusion zone to the width of the weld seam s/b [14]. This means that at a given beam current I_S, a weld seam can only be made narrower by altering the focal distance A_F if, ignoring the electron optical aberrations, the product of the solid angle of the beam and the area of the focal spot of the beam $\omega \cdot F_{Fo}$ becomes smaller. In this respect, even increasing the beam current will only serve to alter the profile of the fusion zone as long as the product $\omega \cdot F_{Fo}$ also does not increase.

These references are intended only to explain the term brightness and its meaning in terms of welding practice. On no account are they intended to imply that a user should modify the electron gun to permit narrower and deeper seams to be welded. Today, one can assume that on the one hand modern electron beam welding machines are already equipped with optimised beam systems, and on the other hand that, for obvious technical manufacturing reasons, the width of the weld seam should not be allowed to fall below a given minimum value.

2.10 Power density

2.10.1 General

In striving to improve both the consistency of the quality of the weld and the setting parameters of different electron beam welding machines, to improve the transfer of welding parameters obtained in trials into actual manufacturing, the varying power across the beam density has been promoted recently as an additional characteristic of interest. Taking total power density as the total beam power P over the area of focus F_{Fo} is not sufficient to characterize an electron beam. Whilst the brightness mentioned in section 2.9 essentially affects the shape of the fusion zone, the entire welding process would seem to be closely related to variation in the power density. More extensive and systematic investigations are required, however, to enable the relationships to be clarified and more efficient measuring instruments need to be developed for use in practice to enable this knowledge to have an economic use in welding manufacturing. The following sections are thus intended to give an overview of the various methods of measurement and to explain some of the factors which affect power density distribution.

2.10.2 Methods of measurement

Determination of the power density distribution requires a measurement of the current density distribution in the electron beam, because the accelerating voltage U_B can be assumed to be constant. The individual methods of measurement differ only in so far as, in the absence of a solid conductor, they use various different aids to measure the current. During the measuring process these aids are subjected to considerable thermal influence by the electron beam, and thus all of the methods used attempt to avoid damaging the sensor (melting) by oscillating the electron beam extremely rapidly.

Purely mathematically, in the ideal case the current density should have the form of a Gaussian or bell shaped curve (Fig. 21). Although almost all of the distributions measured differ to a greater or lesser extent from this normal distribution, it is used to define the beam diameter d_{St}, but not always in the same way. In the literature the definitions given below can be found. In most cases they are based upon mathematical magnitudes and can thus be easily calculated making use of today's advanced data processing facilities, but they are often unrelated to actual welding practice:

d_{St1} = The full width half mean value, determined from the width of the normal distribution at half the value of the maximum power density (Fig. 21).

d_{St2} = Determined from the integral curve (Fig. 22), and characterises a circular surface, (calculated from $(1-1/e)I_S \approx 0.63I_S$) which corresponds to about 63% of the total beam power, P.

d_{St3} = Determined from the integral curve (Fig. 22), and characterises a circular surface, (calculated from $(1-1/e^2)I_S \approx 0.86I_S$) which corresponds to about 86% of the total beam power, P.

d_{St4} = Determined from the integral curve (Fig. 22), and characterises a circular surface which corresponds to about 90% of the total beam power, P.

d_{St4} = Determined from the standard deviation σ of the individual ordinate values in Fig. 21, which is simultaneously the abscissa of the inversion point of the normal distribution.

Comparing measured data is very difficult with these different definitions of the beam and thus of the focal diameter. However, considering the importance of an accurate characterisation of the beam geometry, it is suggested that the

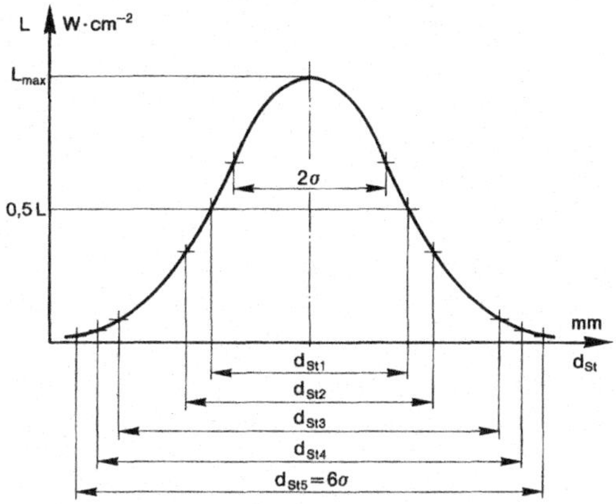

Fig. 21 Assumed normal distribution (Gaussian) for the power density in the electron beam for various defined beam diameters.

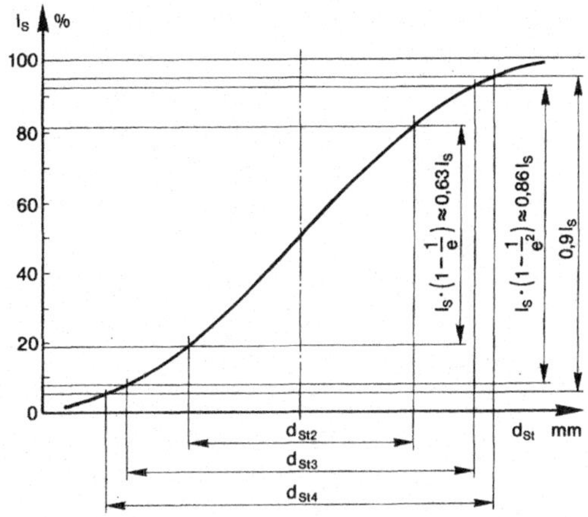

Fig. 22 Integral curve of the normal distribution shown in Fig. 21.

cross section of the beam is considered to be a circular area which includes 90% of the total power of the beam.

Determination of this circular area and its diameter does not involve a great deal of work using today's computer generated current density measurements.

The AB-test[8] is a welding procedure test which requires the simplest preparation of all of the possible methods of measurement [17]. A steel specimen with a number of slots of given dimensions is set up at an angle to the horizontal and welded blind with the electron beam (Fig. 23). Depending upon

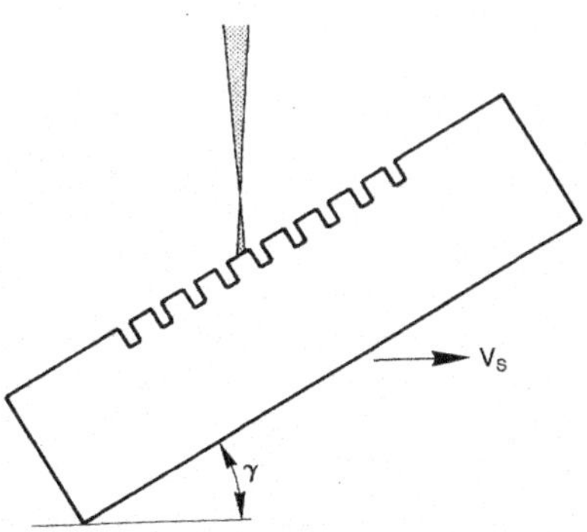

Fig. 23 Diagram of the AB test, v_S = the direction of welding.

[8] AB = Arata Beam

the position of the focus, fusion beads and cutting lines of differing widths occur in the slots, which can be evaluated to provide a measurement of the effective beam diameter. In this way an approximation of the focal distance, A_F, the focal diameter, d_{Fo}, and the beam caustic can be determined. The AB-test is an indirect method of limited accuracy for integrally measuring the beam geometry, which permits no separate information on the effects of the welding parameters.

The threshold method is one of the direct methods of measuring the beam. The electron beam is deflected at high speed over an edge whilst a Faraday cage located beneath the edge absorbs the varying current of the residual part of the beam and conducts this away to be measured (Fig. 24). A further arrangement with sensors at a particular distance serves to determine the speed of oscillation of the beam.

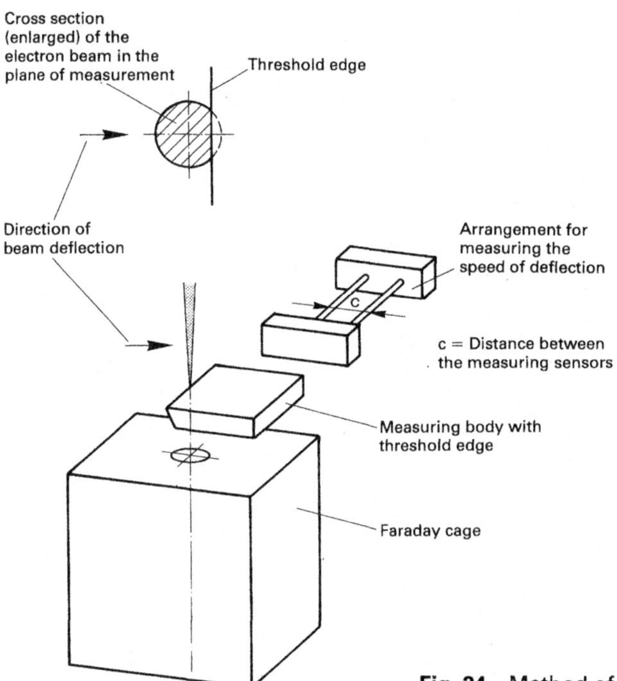

Fig. 24 Method of measuring the threshold value.

The result of this method of measurement is an integral curve of beam current (Fig. 22) from which, after differentiating the current density distribution and with the aid of a chosen threshold value as has already been mentioned ($1-1/e$, $1-1/e^2$, etc), the size of a defined beam diameter can be calculated (Fig. 21). However, the usefulness of the threshold method is limited to the few examples of electron beams having Gaussian shaped and rotationally symmetrical current density distributions.

A similar method makes use of a slit aperture, linearly sampling the rapidly moved electron beam which passes between the measuring edges (Fig. 25).

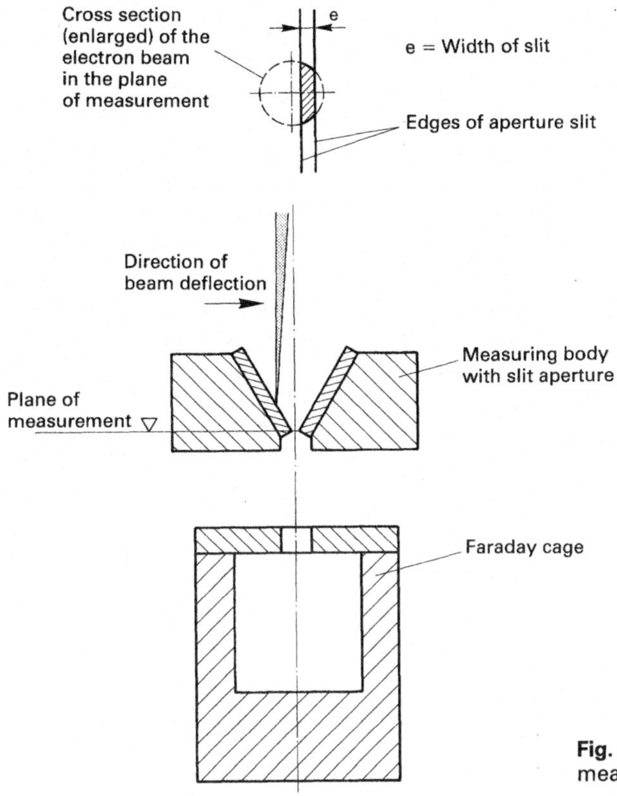

Fig. 25 The slit aperture method of measurement.

After computerised evaluation, the partial beam current measured by the Faraday cage provides a value for the radial current density distribution and a defined beam diameter, whereby it must also again be assumed that a Gaussian distribution exists within the beam and that it is radial symmetrical [18]. If, however, the electron beam is sampled at different angles over an incrementally rotated edge, the axis of rotation being the same as that of the undeflected beam, then it is possible with the aid of computer tomography to three dimensionally plot the current density distribution of any cross section of the beam [19]. The disadvantage of this method lies in the long time required to take the measurements, to process the results and to print out the diagram. Also among the sampling methods are those involving taking measurements with one or a number of wires passing through the cross section of the beam [20].

The wires permit the cross section of the beam to be measured in particularly small elements of area, thus making it possible to measure beams having a Gaussian current density, as well as those systems operating with linear apertures. All of these methods of measurement have in common that they are limited to measuring beam deflection velocities of between 30 and $150\,m \cdot s^{-1}$ and beam powers of 2–15 kW. For measurements of beam powers up to 100 kW and higher, special arrangements have since been developed which are able to deflect the beam at speeds of up to 800 m · s^{-1} [21]. For these methods apertures of molybdenum with diameters of 0.01–0.02 mm are used. These apertures are placed at the position of impingement of the undeflected electron beam in order to ensure that the measurements made are free of distortion. The electron beam, up to 2 mm diameter in the case of high powers, moves in lines extremely quickly across the aperture (Fig. 26), and in this way its entire cross section can be sampled, whilst in the pauses between the individual measurements the beam can be directed into a water cooled copper block.

One particular advantage of this method of measurement, in addition to the extremely high deflection speeds employed, is a specially developed computer program for three dimensional display of the power density from any desired angle of observation (Fig. 27) [22].

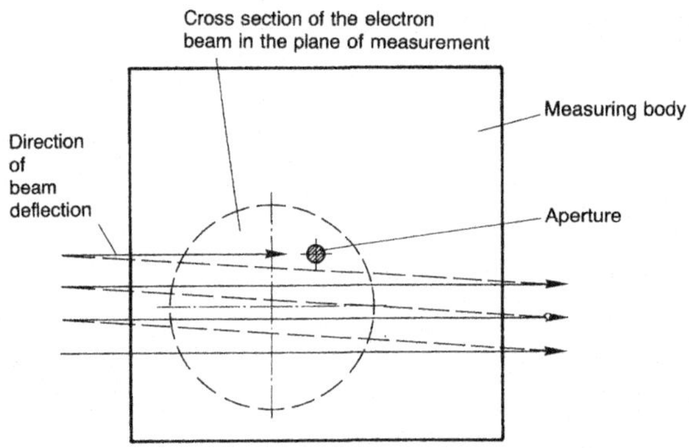

Fig. 26 The aperture method of measurement.

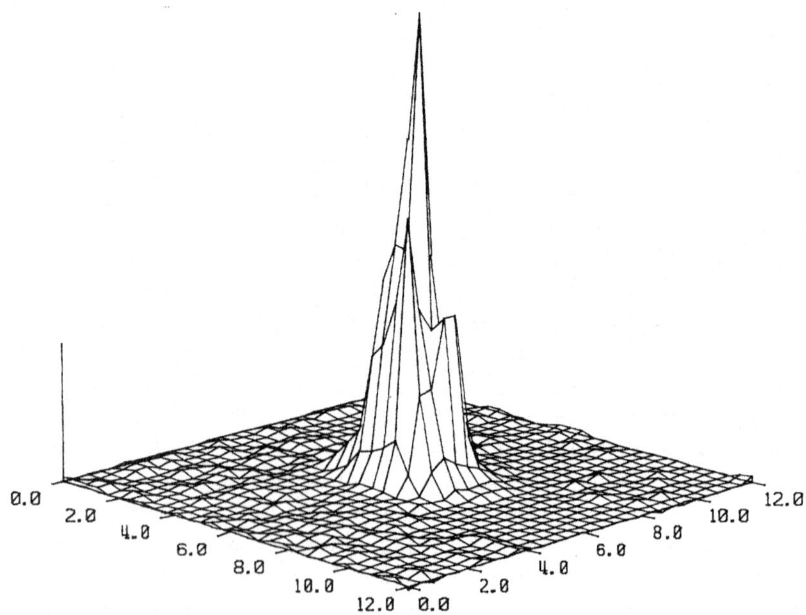

Fig. 27 3-D representation of beam power density with optimum parameter settings.

This type of display increases the clarity of the processed measurements and considerably simplifies evaluation of the different distribution shapes. The X- and Y-axes of the 3-D display are scaled in millimeters and permit a direct determination to be made from the dimensions shown on the distribution representation. The tip of the figure displayed corresponds to the maximum power density, in Fig. 27 this being $L = 35.61\,kW \cdot mm^{-2}$; all other values on the Z-axis are then calculated as a percentage of this. When comparing a number of power density distributions, the 3-D representation may also be made relative to any given Z-co-ordinate. The results may also be printed out in the form of a horizontal cross section through the distribution, in other words as contours of equal power density (Fig. 28). The outer contour lines surround at least 10% of the maximum power density, the next inner contours 30% and so on. As the determined geometry of the beams often deviates from the ideal concentric shape, an additional boundary line is calculated corresponding to a predefined fraction of the beam power of, for example, 90%. The area enclosed by this line is given an 'equivalent beam diameter' corresponding to a circular area of the same content.

2.10.3 Effects on power density distribution

Using this aperture method and modern powerful electronic data processing it is now possible to obtain a simple picture of the power density distribution within the electron beam in a relatively short time. Even though numerous such measurements are already available, the process of introducing this method into actual practice is still very much in the beginning [23]. Figures 27 and 28 show an example of a 3-D plot of a beam with a correctly heated cathode. Even with these apparently optimum parameter settings, the beam power distribution is at best Gaussian and is certainly not rotationally symmetrical. The picture changes drastically, however, if the cathode is under heated (Fig. 29). To maintain the beam current I_S of 160 mA at the same level

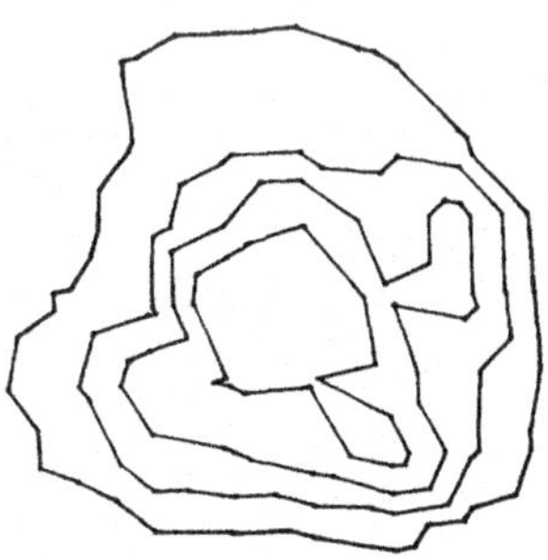

Fig. 28 Contours of equal power density.

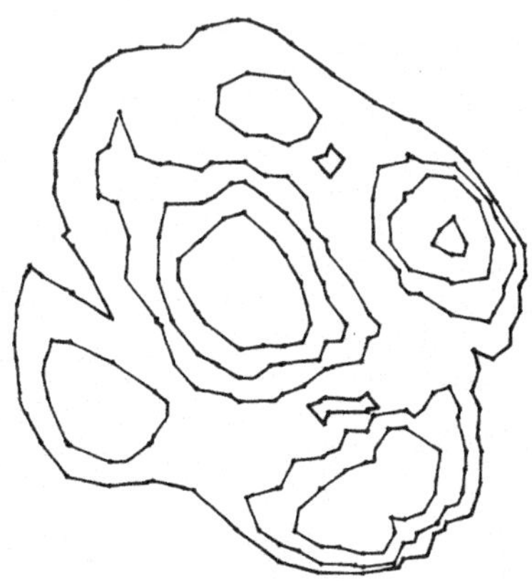

Fig. 29 Contours of equal power density for an under heated cathode.

as previously, the control voltage U_{St} must be reduced excessively and the emitting surface extended over the edges of the cathode. Field peaks are formed at the four corners of the rectangular emitting surface of the cathode, which can be seen in Fig. 29 as subsidiary maxima [23].

The aim of further investigations is to determine the relationship between the shape and quality of the weld as a function of various different power densities.

2.11 Vacuum

For welding, a vacuum in the electron gun is indispensable for three reasons:

- Oxygen must be excluded from contact with the heated cathode, which is at a temperature of about 2750 °C, to prevent it being damaged by oxidation.
- During emission, acceleration and collimation, the light electrons should collide with as few of the heavy molecules present in the air as possible. If they are deflected too far from their ideal path, the electron beam will not be able to be focused to achieve the required power density.
- The field strengths in the electron gun require a high degree of electrical insulation between the anode, the cathode and the control electrode, which can only be ensured by an adequate vacuum. In air or

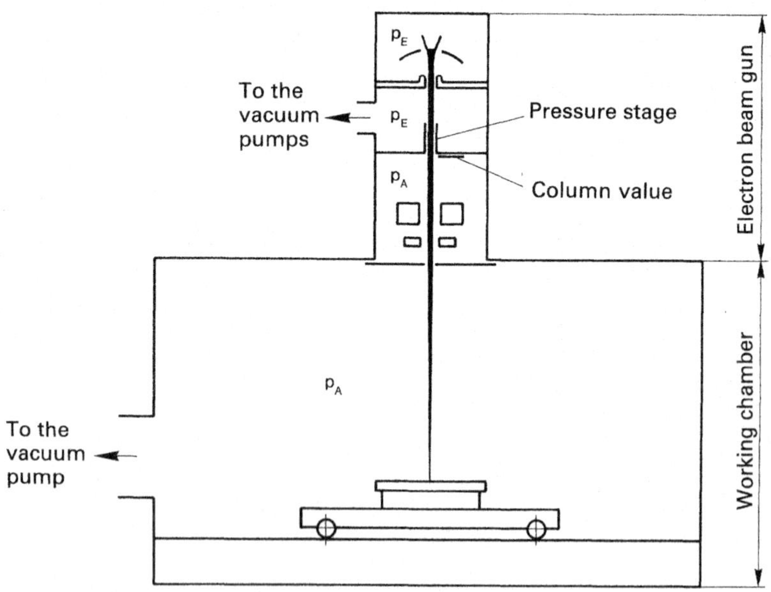

Fig. 30 An electron beam gun and working chamber incorporating pressure stages and column value: p_E = Pressure in the beam generator up to the pressure stage; p_A = Pressure in the working chamber up to the pressure stage.

inadequate vacuum ionisation will take place and the voltages present will arc out.

The highest acceptable pressure within the electron gun is $p_E = 10^{-4}$ mbar and in general in practice evacuation is continued down to a pressure of $< 10^{-4}$ mbar. At a distance from the gun the particularly low pressure required to prevent oxidation of the cathode and to ensure electrical insulation is no longer necessary. With beam paths of for example 500 mm and accelerating voltages of over 120 kV, the pressure can be allowed to increase to about $p_A = 10^{-2}$ mbar before any noticeable scattering of the electron beam by the molecules in the air occurs (see section 4.9). The working pressure at the point of impingement of the beam can thus be from two to three orders of magnitude higher than that in the electron gun whereby in such a case the beam must then be directed through a pressure stage (Fig. 30).

3 The behaviour of the electron beam on penetrating metal

3.1 General

The behaviour of the electron beam on penetrating a metal is one of the most important and difficult parts of the electron beam welding process to understand. Numerous investigations have dealt with the phenomenon by which the beam is able to weld metal several centimetres thick in spite of the fact that the electrons themselves penetrate only a few hundredths of a millimetre into the surface of the material. Even today, some 30 years after the first trials involving welding of greater material thicknesses (see section 1.1), although the physical mechanisms of the electron beam process are known in principle, the individual processes taking part are so complex that they still cannot be quantitively described. The difficulty in accurately and completely describing them lies in part in the fact that the object of investigation, the welding process, to a large extent cannot be studied directly by experiment. Thus, theoretical observations can only be proven by examination of the resulting weld produced. Nevertheless, welding with contrasting materials, radiography during welding, frequency measurements of the penetrating current, as well as calculations of the depth of weld and of the energy processes involved have all served to increase our knowledge of the deep penetration welding effect considerably.

The electron beam welding process is controllable across a very wide range of parameters, as demonstrated by the infinite number of high quality welds produced in many branches of industry. However, this assumes a knowledge of the most important physical relationships and a great deal of practical experience.

3.2 Processes occurring at the surface of the material

As has already been mentioned, the electrons of the focused beam impact with the surface of the workpiece at extremely high speed. How the electrons in the beam transfer their kinetic energy to the workpiece is very complicated and can only be explained briefly here.

Warming of a solid metal results in an increase in the kinetic energy of the atoms in the crystal lattice. Because of their very low mass, electrons are virtually incapable of transferring their energy directly to the considerably heavier nuclei of the lattice atoms in collision processes. It is more likely that the electrons in the beam collide with the electrons in the atomic shell, and in particular with those of the outermost shells and to a certain extent with the free conduction electrons, whose electric fields extend far into their surroundings because of their low degree of bonding.

After this first step in energy transfer, the conduction electrons pass on the collision energy gained [24] to the lattice, again by collision, but this time with the lattice phonons[9)], leading to the intended increase in energy of the lattice atoms. The visible result of this is that the temperature of the metal increases, given sufficient power density in the beam, to far above the boiling point of all known metals. In the liquid and vapour phases, although there is no lattice structure, the energy supplied is nevertheless also transferred in a similar way by collision processes.

Before the path of the electrons through the material is followed further, a whole series of other processes which take place at the point of impingement of the beam must be mentioned. Not all of the electrons in the beam transform their kinetic energy into an increase in temperature of the material (Fig. 31). A small proportion of the electrons, also known as primary electrons, are either reflected elastically directly back from the surface of the workpiece or, after penetrating the surface, are diffused and then reflected back with a partial loss of kinetic energy. Furthermore, in collisions, conducting electrons are ejected from the metal lattice also leaving the material at the surface of the workpiece as secondary electrons.

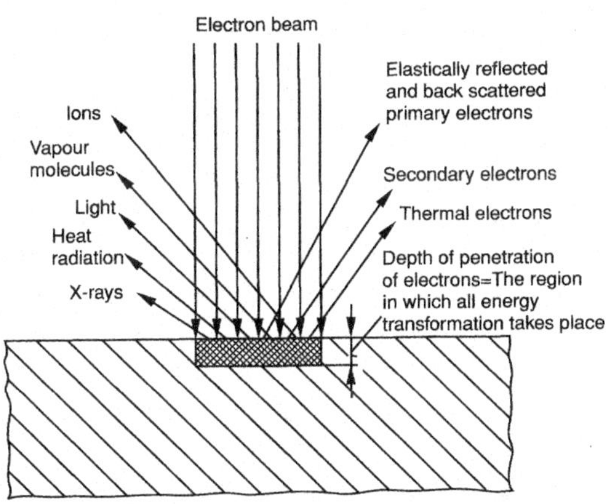

Fig. 31 Processes occurring when the electron beam meets the surface of the workpiece.

The amount, energy spectrum and directional distribution of the back-scattered primary and secondary electrons leaving the surface of the workpiece depend to a large extent on the shape of the workpiece (seam geometry, gap width, edge misalignment, etc) and of the weld pool and vapour capillary, such that they can be used as a physically measurable parameter for weld seam sensing systems and beam focusing aids (see 7.3 and 7.5) [25, 26].

Among other less important back-scattering effects, X-rays are also emitted at the point of impingement of the electron beam. This emission is made up of

[9)] Lattice phonons = Quantised lattice vibrations

X-rays produced as the primary electrons are decelerated, the so-called X-ray continuous radiation, and an X-ray emission characteristic of the material being welded. X-rays have biological effects and in the quantity and energy emitted during electron beam welding are hazardous to the human body. In spite of this, provided that the correct safety procedures for electron beam welding are followed, there should be no danger to the health of the operating personnel. Since welding is normally carried out in a vacuum the wall of the vacuum chamber itself ($U_B \leqslant 60\,kV$), or the chamber wall together with additional protective lead shielding ($U_B > 60\,kV$), will provide adequate and safe protection against X-rays. For more on this aspect see section 11.4.

The sum of all the back scattered losses is dependent upon a number of factors which have been given various levels of importance in the literature [11, 27–29]. The efficiency of the energy transfer process is certainly dependent mainly upon the base material and the geometry of the point of impingement of the electron beam on the surface of the material. From this point of view, the efficiency for steel at the start of welding, that is on a flat target area, is about 0.6, and during welding, once the vapour capillary or 'keyhole' has been formed, 0.9 to 0.95 [10].

3.3 Deep penetration welding effect

From calculations and measurements it can be shown that even after a surprisingly short distance, the beam electrons will have given up their energy to the material. The depth of penetration or the range of electrons in solid steel, for example at $U_B = 150$ kV, is only 0.06 mm (Fig. 32). How then is this value to be related to the known fusion zone which is more than three magnitudes larger, even taking into account the fact that only at the start of welding does the electron beam impinge upon material in either the solid or liquid state, and that only later when in the vapour state does the electron beam have a range that is considerably greater? This small range, practically only encompassing

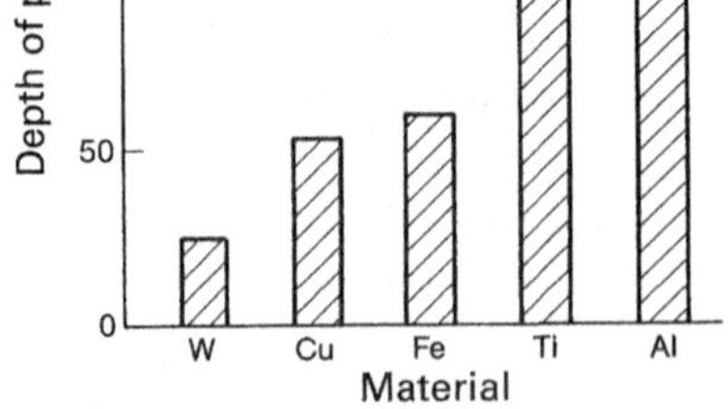

Fig. 32 Depth of penetration of the electrons calculated according to the Schonland formula for a beam of $U_B = 150\,kV$ impinging upon different materials and which has expended ca. 99% of its kinetic energy.

the surface, is paradoxically actually the requirement for being able to weld material several centimetres thick.

In contrast to autogenous gas and arc welding which transfer their thermal energy to the metal from the outside of the material, the 'cold' electron beam transforms its energy into heat only once actually inside the material. This alone is not of particular relevance as far as welding goes. However, if welding is carried out with a low beam current or if the power density of the electron beam is reduced to that of a gas flame or electric arc (Table 1) by defocusing, then nevertheless a quite useful fusion zone of 1 to 2 mm width will be formed having the usual broad shallow shape (Fig. 33). It is thus quite possible to use the electron beam to melt material without actually making use of the deep welding effect. In other words it is quite possible to produce joints, if only small depths of welding are required, by making use only of the normal conductive method of heat transfer. At power densities over $10^5\,W \cdot cm^{-2}$ the electrons transform their kinetic energy so quickly into heat within the extremely short range already mentioned that the material at the point of impingement of the electron beam initially melts without conducting any significant amount of heat into the adjacent regions of the material. The molten material is then further heated, it vaporises, and the vapour is then in turn superheated.

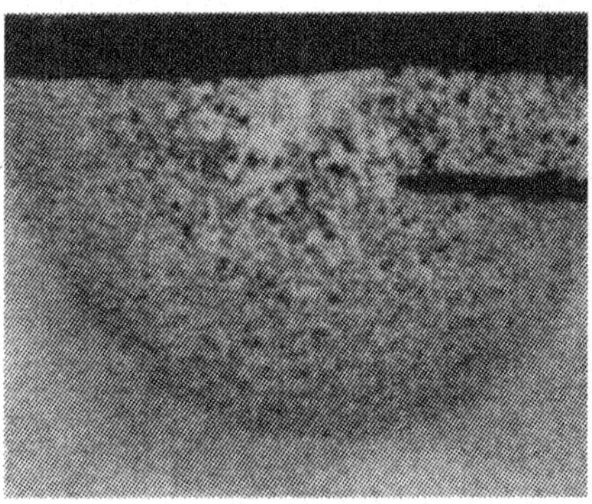

Fig. 33 Macroscopic cross-section through a circumferential weld in Zircalloy 2, showing the fusion and heat affected zones and absence of deep penetration welding (25:1, reproduced 0.6 X).

At temperatures of about 2700 K and above [11], the vapour expands, is released upwards and as a result of the reaction force presses the melt downwards and away sideways.

This forms a depression which permits the electrons to impinge upon fresh material which is heated in the same way. The depression becomes larger and transforms into a keyhole the centre of which consists of vapour surrounded by an envelope of molten material (Fig. 34). Given sufficient energy input this keyhole will penetrate through the entire thickness of the workpiece even if this is as much as several centimetres (Fig. 35). In this way, as in the case of every other welding process, the faces of at least two component parts can be melted, the molten material flows back into the molten pool and the relative

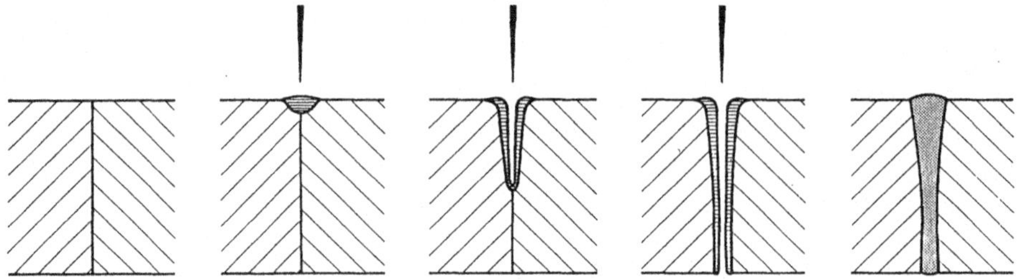

Fig. 34 The various stages of deep penetration welding.

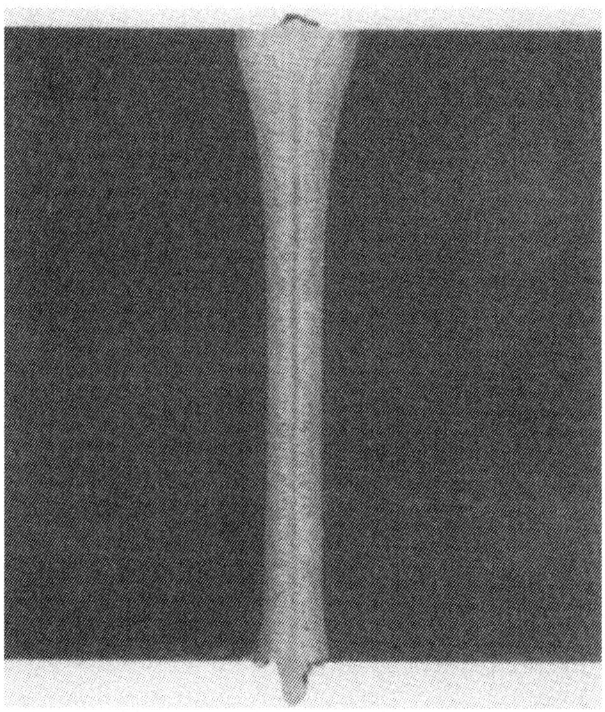

Fig. 35 Macroscopic cross section through the fusion and heat affected zones of a 50 mm thick longitudinal weld in X 20 Cr 13.

movement between the beam and the workpiece solidifies the molten pool to form a welded joint.

The deep welding effect is a physical process in which the electron beam is able to penetrate into the material, far beyond the minute range of the electrons, by forming a keyhole of vaporised and molten material. For this keyhole to remain open continually to permit the transfer of energy necessary for welding to take place, various forces must interact (Fig. 36). The vapour presses from inside outwards on to the wall of the keyhole, and thus also acts against the hydrostatic pressure and surface tension of the molten envelope.

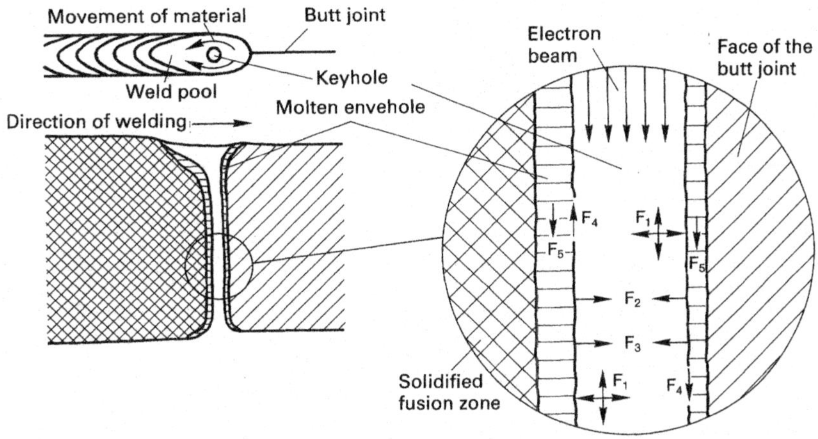

F_1 = Vapour pressure; F_2 = Force resulting from surface tension; F_3 = Hydrostatic pressure; F_4 = Frictional force from the escaping metal vapour; F_5 = Weight of the molten mantle.

Fig. 36 The major forces acting within the keyhole and molten envelope in deep penetration welding.

The upward – as a result of through welding – and downward flowing vapour applies a force to the inner wall of the molten envelope by friction against which the weight of the molten material reacts, these forces thus supporting the weight of the melt.

In addition to this there is a considerable amount of movement of the melt generated within the keyhole once the electron beam begins to move through the workpiece in the direction of welding. It has been surmised from actual visual observations made of the electron beam welding process, further confirmed by extensive investigations, that the forces and movement which occur during this process are only in a state of quasi-equilibrium [30–37].

At welding speeds of about 5–20 $mm \cdot s^{-1}$ and full penetration depths of about 20 mm, the vapour keyhole and molten envelope formed in materials suitable for welding has the more or less inherently stable shape shown in Fig. 36. At lower speeds of welding, however, or in particular in partial penetration welds in which the fusion zone ends within the thickness of the workpiece, a certain characteristic instability is observed. In these cases the vapour pressure is unable to hold the melt in a stable uniform shape and the molten envelope bulges outwards, collapses and falls to the bottom of the keyhole (Fig. 37). The depth of the keyhole is thus reduced and its geometric dimensions and the physical forces involved are lost. Only by the advancement of the electron beam or of the workpiece can the keyhole be reformed, which then starts the process leading to a fresh collapse of the molten envelope.

As shown in Fig. 37, the collapsing molten envelope can also block off a part of the keyhole. With the rapid solidification occurring, this becomes isolated from the melt so that after the vapour has condensed it can no longer be filled by molten material. A root cavity is thus formed.

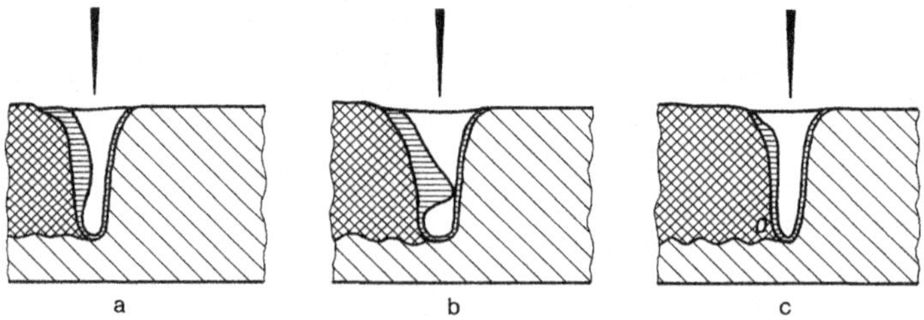

Fig. 37 Formation of a void at the base of the fusion zone as described in [30]: *a)* The trailing part of the molten envelope builds up and collapses; *b)* The lower part of the keyhole becomes cut off and the electron beam is no longer able to reach the bottom of the fusion zone; *c)* The collapsed molten mantle has solidified enclosing a void.

The dimensional scale of the keyhole and shape of the molten envelope shown in Fig. 37 are not typical of all electron beam welds and have only been sketched in this way for clarity. From the example of the longitudinal section through an electron beam weld shown in Fig. 38 it can be seen that the base of the fusion zone is made up of numerous narrow needles of fused metal partly punctuated with cavities. This has been termed 'the spiking effect' or 'root spiking'.

Fig. 38 Longitudinal section through the root of a partial penetration weld showing the spiking effect and entrapment of voids. Material: X 6 CrNiTi 18 10 (15:1, Reproduced: 0.6 X).

In the past a considerable effort has been made to intervene in these dynamic effects using various techniques, with the aim of stabilising the process and preventing the occurrence of defects in the weld. One of the most interesting suggestions made use of a second electron beam [38]. With this additional electron beam inclined at an angle to the vertical, the base of the keyhole can be enlarged against the direction of the welding and the spiking effect prevented. The solidification processes thus progress continuously without formation of any cavities or variation in depth of the fusion zone (Fig. 39).

The technical expense of this type of tandem electron beam welding (TEB welding) is considerable and hardly justifiable in normal manufacturing practice.

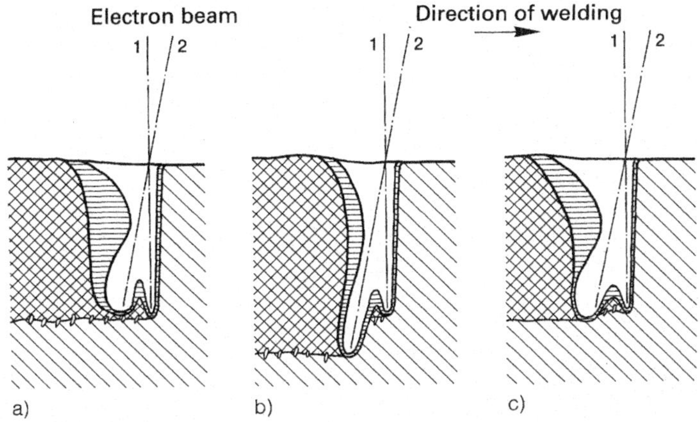

Fig. 39 The effect which the tandem electron beam welding process (TEB) has on the dynamic processes taking place in the keyhole and molten envelope, according to [38]. 1 = Electron beam 1; 2 = Electron beam 2: *a)* Beam power P_2 too low; *b)* Beam power P_2 too high; *c)* Beam power P_2 correctly adjusted to prevent the formation of root voids (spiking).

To improve the weld profile and prevent the undercutting and bulges which occur along the weld cap because of the unstable molten envelope when welding at high speeds, it is helpful to machine a 1 mm wide groove, the depth depending upon the actual speed of welding, as a weld preparation (Fig. 40) [39]. This should only be considered as a special measure, however, as it is much more effective to widen the weld by defocusing or oscillating the electron beam.

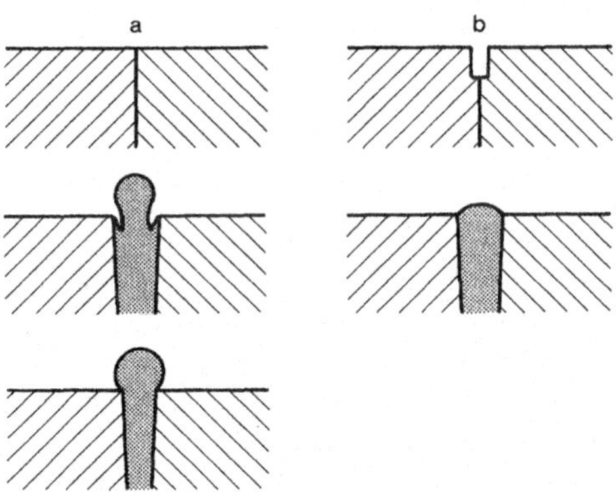

Fig. 40 Welds with excessive camber, undercutting and ridging, and the recommended method of preventing this by slotting the upper part of the butt joint [39]: *a)* Extreme weld camber in the case of a normally prepared butt joint, formed when welding at high speeds; *b)* Weld camber produced when electron beam welding a slotted butt joint.

Further experiments have been carried out to determine whether the dynamic processes taking place within the keyhole result from fluctuations in the energy input such as ripples in the accelerating voltage or the beam current or an irregular welding speed. Under normal conditions of welding, however, no such relationship has been established [34]. In contrast, it is highly likely that the power density has an effect on spiking.

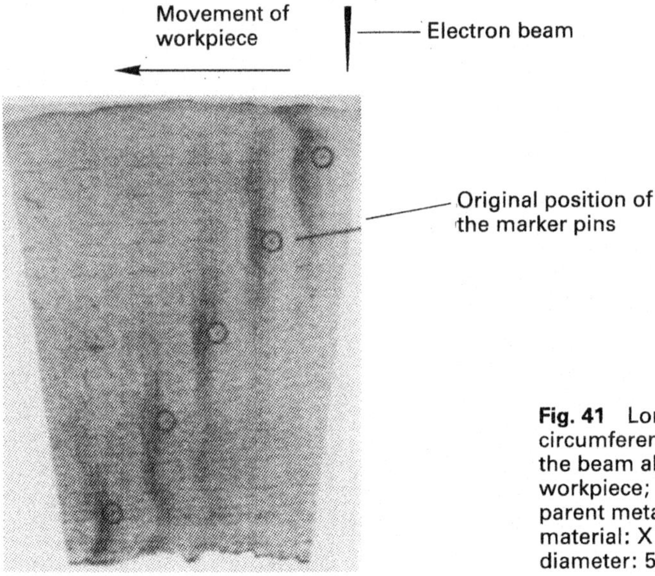

Fig. 41 Longitudinal section through a circumferential pipe weld, welded with the beam almost normal to the workpiece; wall thickness: 100 mm; parent metal: 22 NiMoCr 3 7; marker material: X 6 CrNiTi 18 10. Marker pin diameter: 5 mm [42].

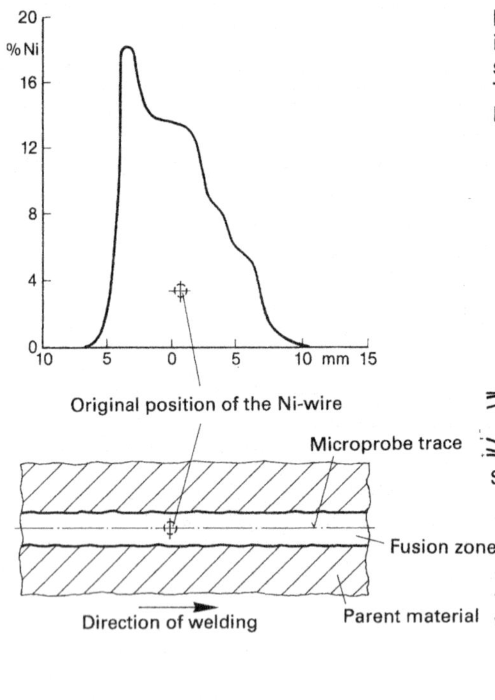

Fig. 42 Distribution of the marker material in a horizontal section 7.5 mm beneath the surface of the workpiece. Parent metal: Titanium, 25 mm thick. Marker material: Nickel. Marker pin diameter: 1.5 mm.

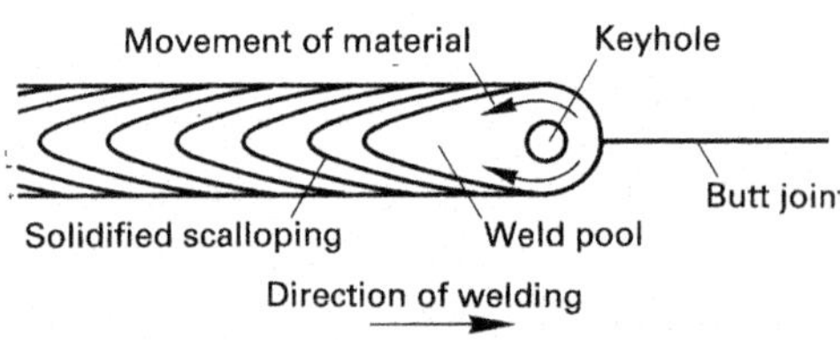

Fig. 43 The movement of material and formation of scalloping on the surface of the weld during electron beam welding.

Amongst other experimental work there have also been investigations into the hydrodynamic processes taking place in the melt [40–42]. In such work use is made of marker materials such as wires, foils and pins placed between the weld faces and then fused together with the base material.

The distribution of the marker material in the fusion zone is then followed metallographically and analytically. This has shown that material moves against the direction of welding in the weld pool as well as in the longitudinal direction of the molten envelope (Fig. 41 and 42). The distribution of the marker material in the longitudinal section (Fig. 41) also provides information about the shape of the keyhole and of the molten envelope at different depths of welding. A further possible method of investigation is to introduce small tungsten balls into the weld pool and to trace their movement radiographically [43].

It is the movement of material in the weld pool and the keyhole, however, that is of greatest practical importance (Fig. 43). This explains the distinct increase in height of the weld cap at the start, and the formation of the crater at the end of the weld seam, both of which can be prevented if suitable precautions are taken (see sections 6.8 and 7.2). It should be mentioned again at this point that the instabilities described occur in particular in the case of welds which do not penetrate through the full thickness of the metal, and when welding thicker walled materials at low speeds. The overwhelming number of electron beam welds made in manufacturing are welds with a root cap in material of average wall thickness made at relatively high speed. Under these conditions any instabilities in the dynamics of the keyhole do not adversely affect the quality of the weld.

3.4 Consequences for the welding process

Many models and calculations have been made of the magnitudes and directions of the forces involved in the deep welding effect, and these will not be discussed further here [44, 45]. The considerable knowledge gained from this work provides very useful information which aids in the correct choice of settings to be made for the welding process to ensure production of highly consistent good quality welds. The position of focus of the beam in relation to the surface of the workpiece together with the various parameters governing oscillation of the beam are of particular importance to the processes occurring during welding and to the shape of the fusion zones formed. This is dealt with in greater detail in section 4.6. Further important factors already mentioned are the beam energy and the distribution of power density within the beam. A quantitative or at least qualitatively formulated rule governing the relationship between the power density distribution and the mechanisms taking place in the welding process has not yet been realised.

From the explanations given regarding the behaviour of electrons on penetrating a material, the following recommendations can be made for welding practice:

- Welds with a root cap (full penetration welds) are always to be preferred.
- If the workpiece is thicker than the required depth of weld then the faces to be welded should be prepared such that the electron beam will nevertheless form a root cap (see section 6.3.3).

- With welds without root caps, the bottom of the fusion zone should be designed to be outside the load carrying cross section of the component (see section 6.3.1).
- Particular settings for oscillation of the electron beam should be determined for the run out regions of circular weld seams to prevent formation of root cavities by specifically and periodically varying the energy distribution [46].
- Over-convex weld capping and the occurrence of end craters in circular welds should be avoided by suitably controlling the beam current.

4 Welding parameters and advice on welding practice

4.1 General

Figure 44 shows the items in the electron beam gun and the working chamber which are used for direct control of the most important electron beam parameters. The following sections not only aim to explain these parameters, but also to give an indication of their respective relationships with the geometry of the weld seam and the avoidance of defects, etc. Their effects on the metallurgical processes involved are explained in section 5. The following diagrams include measurements made on various types of electron beam welding machine and are intended to show the trends in the respective relationships. The numerical values are thus not intended to be typical of all types of machine.

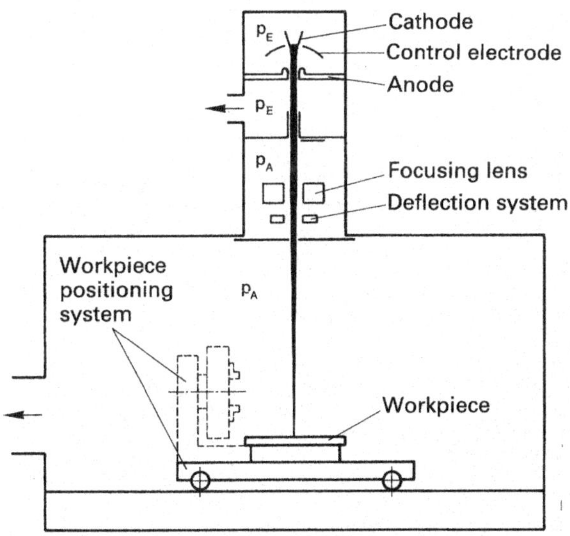

Fig. 44 An electron beam gun and working chamber using various systems and pressures directly to affect the electron beam.

4.2 Accelerating voltage

From the physics point of view, an accelerating voltage is among the most elementary requirements needed to produce a high energy beam of charge carriers to be used for welding materials. But from the welding technological point of view, the accelerating voltage is not the most important of the para-

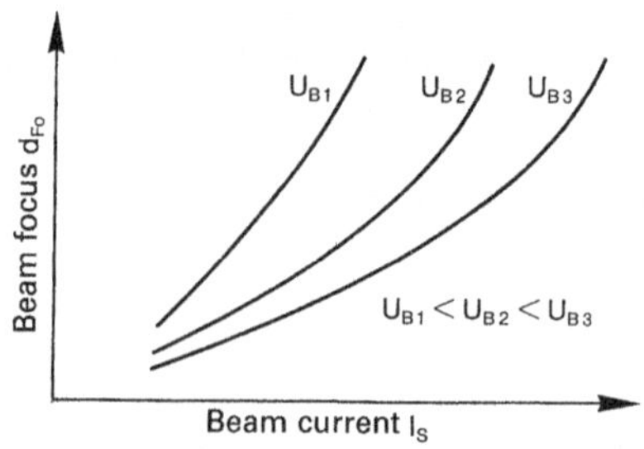

Fig. 45 Effect of beam current I_S and accelerating voltage U_B on the focus diameter d_{Fo} [47].

meters involved. For the majority of welding operations the accelerating voltage is kept constant at U_{Bmax}, which depending upon the type of high voltage generator and electron beam gun, is set at 60 or 150 kV.

In this way, as can be seen in Fig. 45, for the required beam power P and lowest beam current I_S, the smallest focal diameter d_{F0} is achieved for the beam. Only in certain exceptional instances therefore, for example when wide weld seams are to be produced using a larger beam focus, is it appropriate to use lower accelerating voltages.

Figure 46 shows the effect which the accelerating voltage has on the geometry of the weld seam at different focal distances. If A_F is increased from 350 to 1200 mm, then at $U_{B1} = 150$ kV the depth of the fusion zone decreases by about 30% and at $U_{B2} = 60$ kV by about 40%. At the same focal distance A_F, the width of the seam at $U_{B2} = 60$ kV is about 30% greater than at

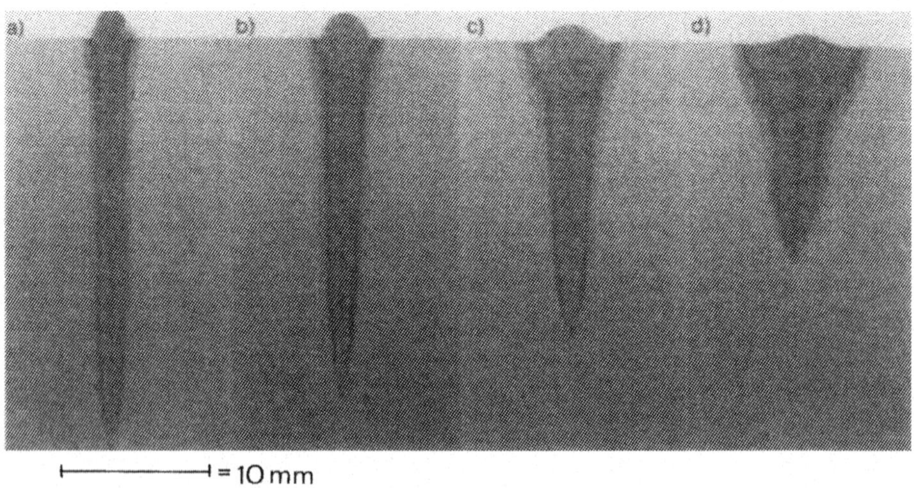

Fig. 46 Cross sections through welds made with different accelerating voltages U_B and focal distances A_F, but with a constant beam power of P = 5 kW. Showing the effect on the depth of the fusion zone s and width of the weld b.

	U_B, kV	A_F, mm
a	150	350
b	60	350
c	150	1200
d	60	1200

U_{B1} = 150 kV, and at a focal distance of A_{F2} = 120 mm the width is in both cases double that of the initial value. From this it can be seen that with an electron beam with a high accelerating voltage, the focal diameter increases less as the accelerating voltage increases than in the case of an electron beam with a low accelerating voltage.

When welding materials for which they or their alloying components have a high vapour pressure, high voltage arc outs can occur. These are electrical discharges which occur when the insulating effect of the vacuum or the electric field in the electron beam gun are interrupted.

On the one hand, the cause can be desorption processes taking place on the surface of the cathode and the high voltage insulators. On the other hand, they can be caused by metal vapour containing molten particles which erupts under the pressure in the keyhole and is projected through the annular anode and into the electron beam gun. Particles of dirt, in particular in the space between the cathode and the annulus of the control electrode are also a factor which can lead to voltage peaks forming in the electrical field. As a result of the electrical discharge which thus occurs, the control voltage collapses, the beam current increases until a current limiting relay cuts out, leading to defects being formed in the weld (voids, weld burn through, etc). Finally, the electron beam is suddenly interrupted, generally leaving behind an end crater. Thus, after such a discharge, defects may be left in the welded workpiece which can only be repaired subsequently with difficulty.

A series of steps has been developed to prevent such defects. Amongst these are welding at very great working distances ($A_A > 2$ m) and use of electrical control systems which prevent an extreme increase in current and which, within only a few milliseconds, reset the most important beam parameters to their initial settings [48, 49].

Another possibility is to deflect or bend the electron beam in such a way as to ensure that the molten particles ejected from the point of impingement of the beam cannot enter the electron gun. Figure 47 shows a magnetic trap which is arranged beneath the anode and which directs the beam around a gate [50]. In the case of manufacturing welding machines, however, it is recommended

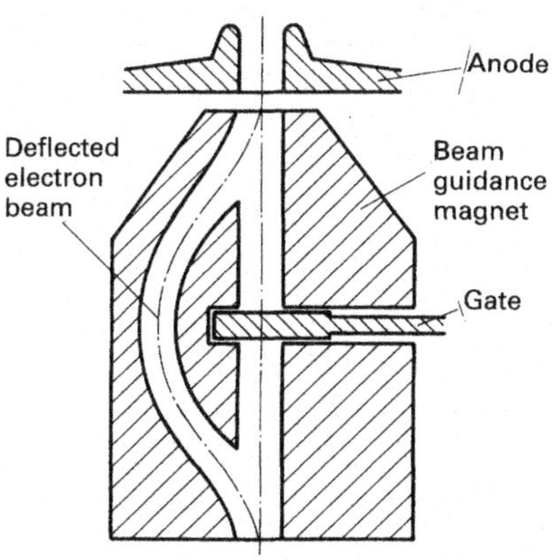

Fig. 47 Magnetic trap beneath the anode to prevent high voltage arc outs in the beam generator caused by returning metal vapour [50].

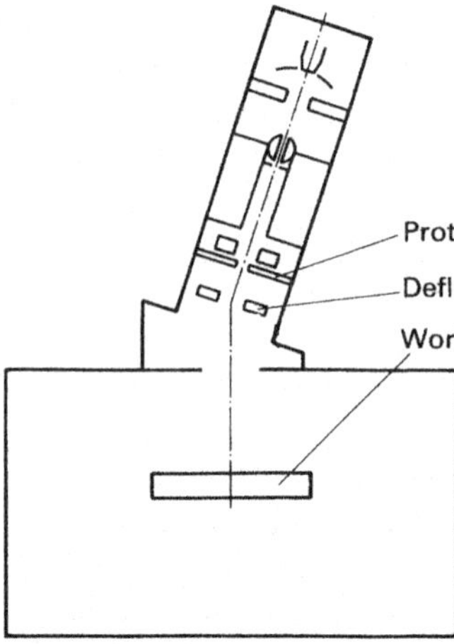

Fig. 48 Beam deflection achieved by tilting the electron gun.

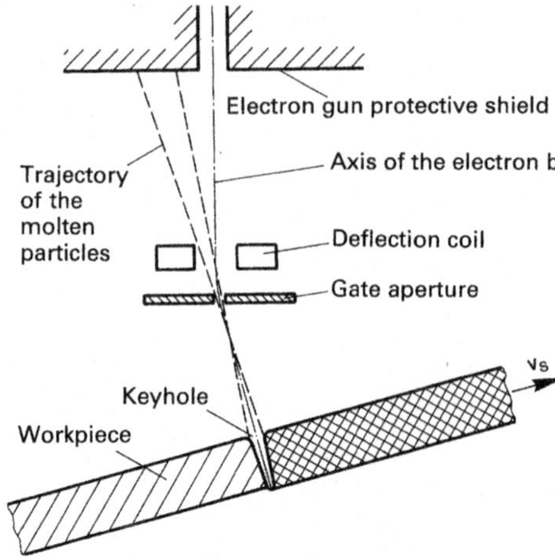

Fig. 49 Beam deflection outside the electron gun.

that the electron beam be bent and that the electron gun be mounted accordingly at an angle above the working chamber to enable the beam to impinge at right angles to the surface of the workpiece (Fig. 48). In some special welding machines it is often more advantageous to bend the electron beam once it has actually entered the working chamber (Fig. 49) in order to avoid the complicated modifications otherwise necessary [51]. The diverging molten particles leave the point of welding in the direction of the longitudinal axis of the keyhole. The angle of the beam must then be such that the particles cannot enter the point of exit of the beam from the electron gun, but impinge upon a suitably positioned shield.

It should not be overlooked that bending the beam beneath the electron gun causes a number of difficulties. For instance, it is necessary to modify the optical path to enable the butt weld preparation to be positioned in accordance with the modified beam axis. In addition, bending the electron beam causes it to become distorted, which then also requires appropriate compensation.

Bending the beam is thus only justified in certain special cases, for example when welding particularly heavy wall thicknesses. With correct handling of the electrodes and provided that the electron gun is kept clean, today's electron beam welding machines may be regarded as short circuit discharge proof for the vast majority of the materials suitable for welding.

4.3 Beam current

The beam current, together with the accelerating voltage, determines the beam power and above all affects the resulting weld. As shown in Fig. 50, it is the impinging beam current which mainly influences the welding process, and which subsequently, providing that the full thickness of the workpiece is penetrated, divides into the workpiece current and the transmitted current. The

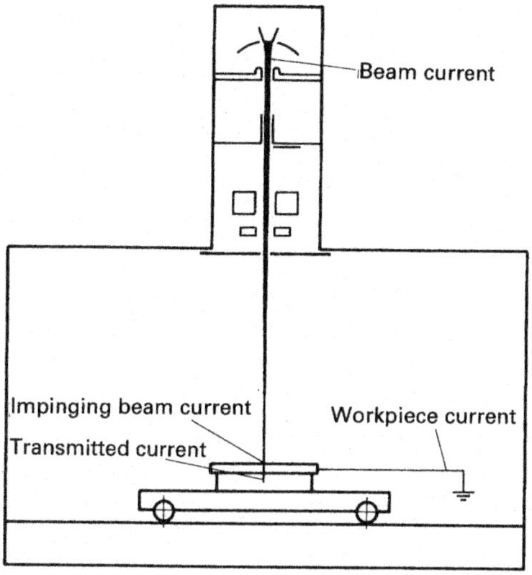

Fig. 50 The beam current, impinging beam current, transmitted current and current induced in the workpiece, in the electron gun and the working chamber.

greater proportion of the beam electrons go to form the workpiece current, which is then conducted to earth through the working table, the clamping arrangement, etc. In contrast the residual energy of the electron beam, in the form of the transmitted current, keeps the keyhole open on the underside of the workpiece and ultimately affects the shape of the root cap of the weld. If an excessively high beam current is set for a particular welding operation, the surface tension developed during the liquefying phase may not be sufficient to support the over large fusion envelope and weld pool against the force of gravity.

In this event the weld metal flows downwards, causing concavity in the crown of the weld and droplets of metal along the root or falling from the root (Fig. 51*a* and *b*). Formation of such droplets represents a loss of material from the weld and is caused by the transmitted current being too high, dropping the molten metal downwards and causing it to form droplets. In contrast, a beam current which is set too low will not completely melt the full thickness of the workpieces to be joined, and in addition will result in an excessively convex weld crown (Fig. 51*c*, also see section 4.10) The 'optimum' beam current completely welds through the full thickness of the workpiece and results in formation of a smoothly convex shaped weld crown and root (Fig. 51*d*).

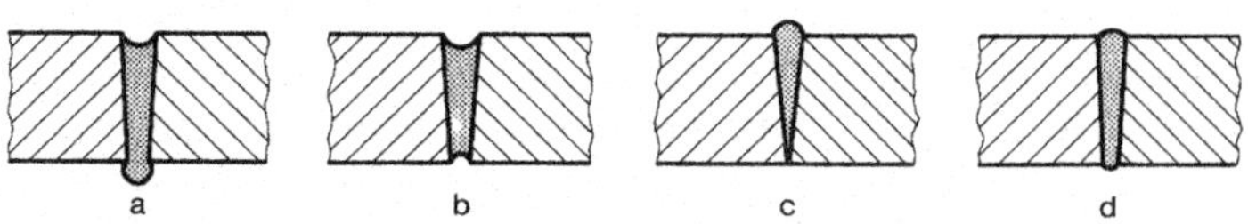

Fig. 51 Different effects achieved in welding a full penetration joint: *a)* Too heavily through welded, concave weld crown and droplets formed along root; *b)* As *a)* but with root sprayed out; *c)* Insufficiently through welded, convex weld crown, root insufficiently welded; *d)* Normal full penetration weld.

Of all the parameters, the beam current most often requires continuous control or adjustment in order to adapt the power of the beam to different welding operations. Modern electron beam welding machines are, or can be, additionally equipped to carry out the following groups of control functions:

- Beam current control, also called 'slope control', which is used to prevent excessive weld convexity and the occurrence of end craters in circular welds.
- Beam current control for welding materials with varying wall thicknesses. This is achieved by using a control system based on beam path length.
- Measurement of the transmitted current in order to control the beam current, e.g. when welding heavier wall thicknesses at lower speeds. In this case the temperature of the workpiece increases considerably during welding, which requires a suitable reduction of the beam current.

With these types of welding operation the accelerating voltage and welding speed are generally not controlled, see also section 7.2.

4.4 Lens current, focal position

The electron beam diverges on leaving the electron gun and is collimated and focused by an electromagnetic lens to achieve its typically high power density. To focus the beam accurately on the surface of the workpiece with the aid of the lens current to facilitate welding requires a number of additional pieces of equipment. In the simplest case, before welding, the electron beam is set up on a piece of material set at the same height as the actual workpiece (Fig. 52).

By altering the lens current and observing the point of impingement through an optical system, the smallest beam diameter can be set and welding begun. Actual experience, however, shows that defect free welds, particularly in the case of very deep welds, can only be achieved with difficulty using this focal position. The beam current range between the workpiece 'not being welded through the full thickness' and 'excessively over-welded' is very small and either the crown of the weld will be seen to exhibit undercutting or the root to be covered with droplets. If the electron beam is set to be de-focused, however, that is with the position of focus either above or below the surface of the workpiece (Fig. 53), then the occurrence of these defects can be considerably reduced. De-focusing is a simple yet very effective way of affecting the relationship between the forces in the keyhole and the molten metal envelope on the one hand, and the force of gravity on the other. By how much the beam must be de-focused – expressed in mA of lens current or mm of focal shift – can only be determined using trial welds because of the numerous other factors involved. Figure 54 shows a method which can be used to show the effect of focus position on the weld. In the case of steels, with thicker welds of for example $t > 10$ mm, a focus position of about $0.3t$ beneath the surface of the material has been shown to be most suitable.

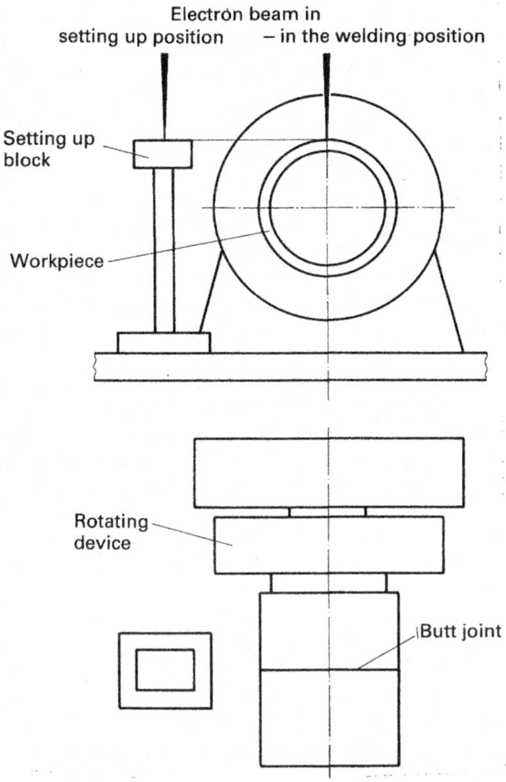

Fig. 52 Showing how to set the focus for a circumferential weld.

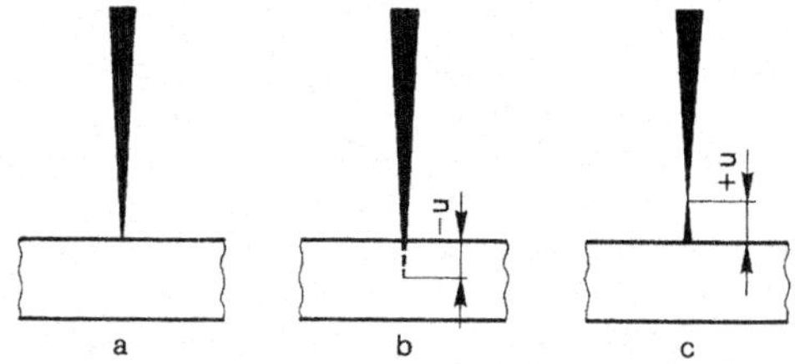

Fig. 53 Various focus positions:
a) Normally focused; *b)* Under focused;
c) Over focused.

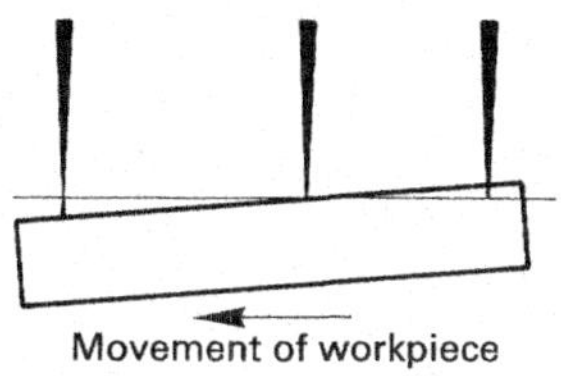

Fig. 54 An experimental way of welding with a continuously changing focus position.

Experience has thus shown that an electron beam focused to the smallest possible diameter at the point of impingement is certainly not the most suitable for all welding operations. Depending upon the type of material, the speed of welding, the weld gap, etc, in almost all cases it becomes necessary to de-focus and – as explained in section 4.6 – to oscillate the beam in order to suppress the occurrence not only of external but also internal defects (shrinkage cavities, pores), or at least to reduce them to a minimum.

In respect of beam focusing the terms 'normal' and 'optimum' focus are often used with no firm definition or agreed terms of use. In welding practice it is most useful to use the term normal focusing of an electron beam to mean setting the focus of the beam on the surface of the workpiece (thus also referred to as surface focus) such that the smallest possible focal spot diameter is achieved. Optimum focusing is thus in contrast applied to a beam focused to produce the fewest possible defects in the weld. In this case the beam impinges on to the surface of the workpiece with a spot diameter greater than the actual focal diameter (Fig. 55 and 56).

Without taking into account the required focal position, the normal lens current for the given accelerating voltage and focal distance must be determined,

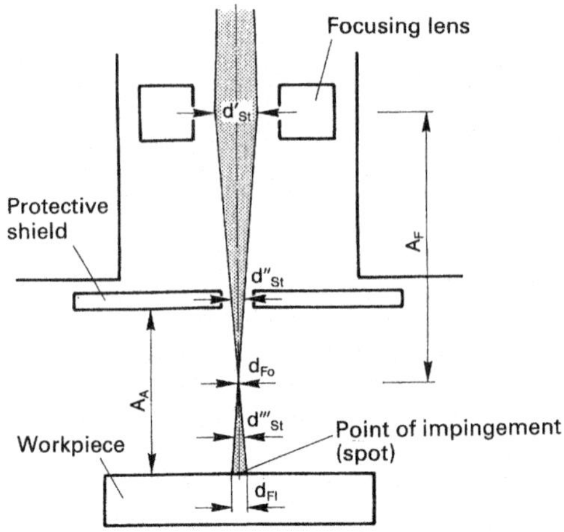

Fig. 55 Definition of focal diameter d_{Fo}, spot diameter d_{Fl} and beam diameter d_{St} in various planes of measurement, working distance A_A and focal distance A_F.

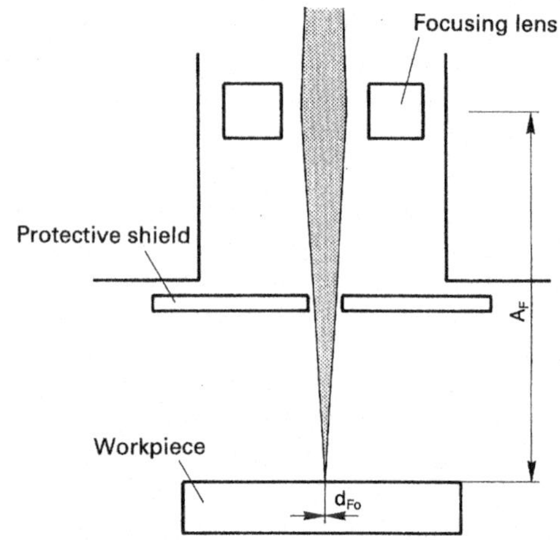

Fig. 56 Distance A_F under 'normal' focusing conditions.

as has already been mentioned, by observing the diameter of the spot on a testpiece.

On high beam currents this leads to considerable difficulties, because during the setting time the whole beam power penetrates the testpiece and leaves a crater at the point of impingement. The impinging beam will be visible as a bright spot on the workpiece surface but a focus position beneath the workpiece surface is undefined. Practically, at beam powers up to 10 kW, these difficulties are avoided by splitting parameter setting into two steps. First with a beam current suitable for the testpiece, lens current for 'normal' focus is determined. This lens current, within the power range, is sufficiently accurate as a reference for focusing the final beam current and may be used to determine the optimum focus for welding.

To determine the optimum focus position for high beam currents it is also possible to proceed in accordance with Fig. 54. For completeness methods for automatic focusing should also be mentioned. But this needs a large investment in additional equipment [52–57], see section 7.3. Indeed, users must wait for these attachments, because they have not yet found their way into manufacturing welding machines. Defocusing of the electron beam is not only, as has already been mentioned, of great importance to the shape of weld bead produced, but also to the depth of the fusion zone and thus the maximum thickness of seam which can ultimately be welded. As an example, Fig. 57 shows the effect which the lens current has on the depth of the fusion zone at various beam currents. Whilst at $I_S = 20\,mA$, the depth of the fusion zone varies rapidly with the lens current I_L, at higher beam currents the depth of the fusion zone exhibits less sharp variation. Here, at the same time, the maximum depth of fusion zone is achieved not at normal lens currents, but with slight

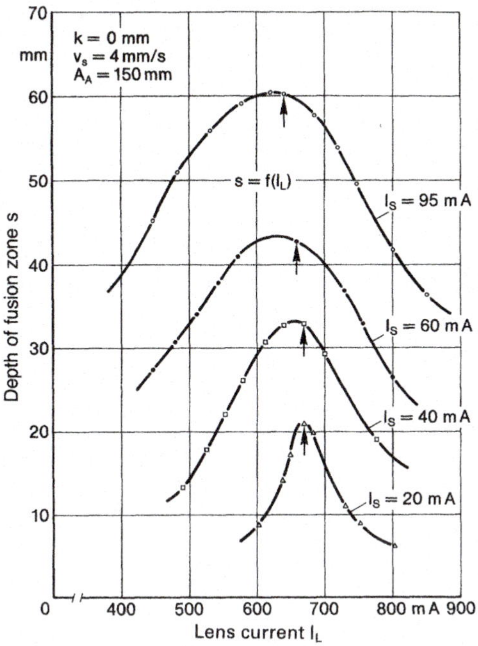

Fig. 57 Depth of the fusion zone s as a function of the lens current I_L, k = the vertical distance between the cathode and the control electrode. Material: Ck 22, ↑ = Normal lens current [58].

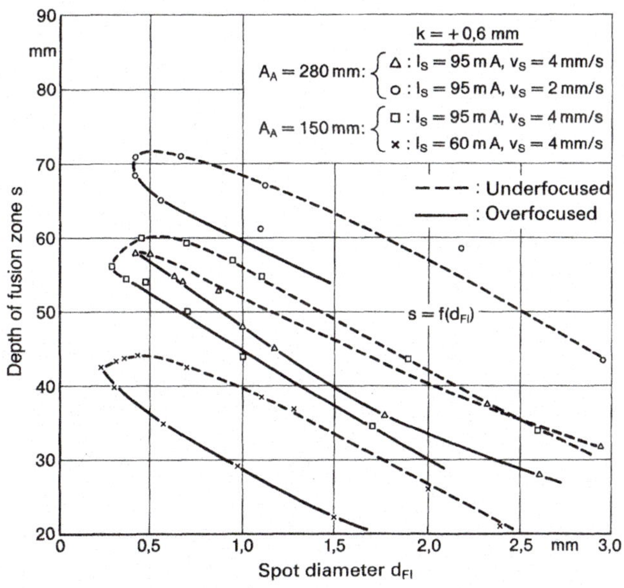

Fig. 58 Depth of the fusion zone s as a function of the spot diameter d_{Fl}, otherwise all other details as in Fig. 57.

under-focusing. Figure 58 emphasises this relationship and illustrates the relationship between the depth of the fusion zone and the spot diameter. It is surprising that for the same beam diameter at the surface of the workpiece, but on the one hand with a diverging electron beam (over-focused) and on the

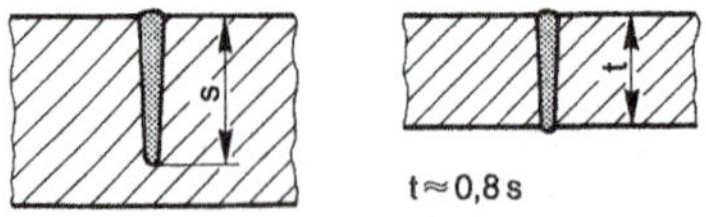

Fig. 59 The relationship between the depth of the fusion zone and the thickness of the weld for constant welding parameters.

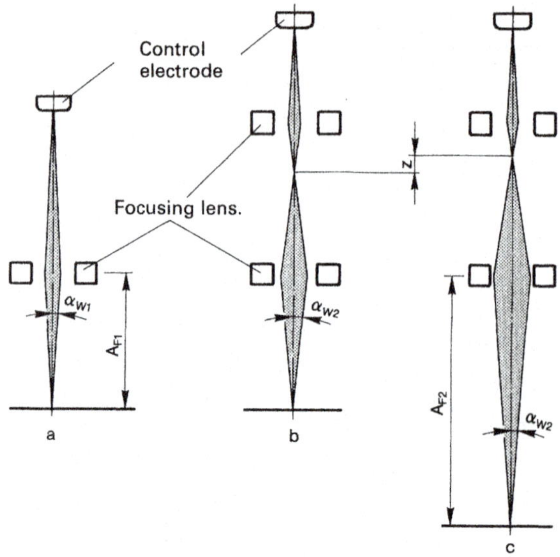

Fig. 60 Effect of different focusing systems on the electron beam: *a*) Normal focusing system; *b*) Enlargement of the aperture angle by a second focusing lens at the same focal distance as in *a*); *c*) By a second focusing lens of the same aperture angle but at a greater focal distance compared with *a*).

other with a converging beam (under focused), differences in the depth of the fusion zone s of more than 10 mm can occur.

Figures 57 and 58 show the depth of the fusion zone s as a function of different parameters. The depth of the fusion zone is not to be compared with the thickness t of an I-shaped weld bead (Fig. 59).

For welding an I-seam with a root, the electron beam must pass through the workpiece with a certain residual power, which in the case of a weld bead without a root (i.e. not passing through the full thickness of the material) would be available to increase the depth of the fusion zone. Thus with the same settings, when welding through the full thickness of a material, the depth of fusion is limited to about 80% of that which would be achieved by a weld only partially penetrating the full thickness of the material.

When electron beam welding with increasing focal and working distances, the focal diameter increases whilst the maximum depth of fusion decreases (Fig. 46). This decrease in the power density of the beam can be countered by use of a second focusing lens (Fig. 60). This lens can either be built into the gun itself or be mounted in the working chamber. This permits the aperture angle α_w to be increased and the focal diameter again reduced. According to the rule of constancy of the brightness, R, the power density increases proportionally with the spacial angle ω or with the square of the aperture angle α. This method of controlling the beam characteristics is of particular practical importance when welding thicker workpieces [42].

4.5 Welding speed

With the movement of the workpiece relative to the electron beam, or vice versa, the keyhole moves together with the molten metal envelope through the

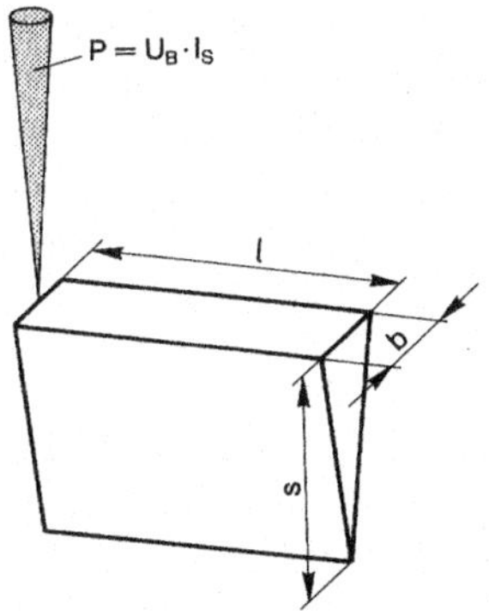

Fig. 61 The volume of material melted by an electron beam.

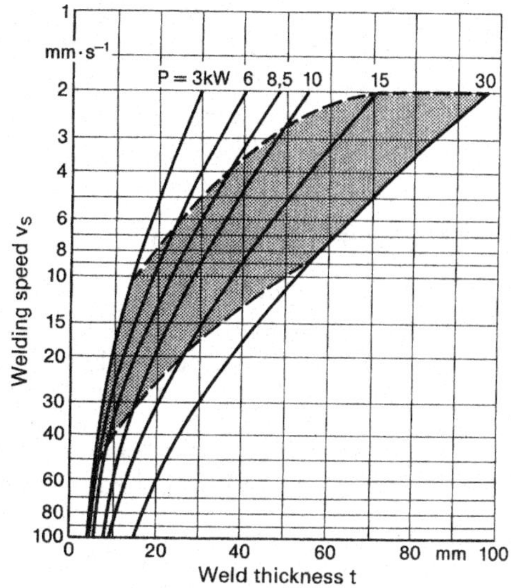

Fig. 62 Guide values for the thickness of welds which can be achieved in steel as a function of beam power and welding speed. P = the maximum available power of the welding machine. Shaded area = the economically viable area of working.

material and a weld bead is produced. As a result of this movement, the power of the beam is transferred into the volume of the material such that for each welding operation a typical geometry of heat input is set up (Fig. 61). In electron beam welding in comparison with other fusion processes the heated volume is particularly small. Investigations have shown that the heat input from a constant beam power is, as would be expected, dependent upon the material and not on either the width of the weld bead or the depth of fusion. With increasing welding speed the heat input decreases as a result of lower loss of heat by conduction [59].

All of the welding parameters mentioned so far affect the welding process quite strongly and in characteristic ways. The beam power together with the focal position mainly affects forces acting in the keyhole and the molten envelope. The dynamic processes, including solidification and cooling, are in the main affected by the speed of welding. The speed of welding together with beam power both determine the depth of the fusion zone or seam thickness (Fig. 62) as well as the transport of material in the weld pool and molten envelope.

Thus at particularly high welding speeds, the thickness of the molten envelope surrounding the keyhole can decrease, which upsets the uniform transport of material within the weld pool and leads to formation of drop-like contractions along the crown and root of the weld (Fig. 40). Widening the weld bead by defocusing or oscillating the beam helps to prevent such irregularities. In contrast, at very low welding speeds an over-wide weld bead may be formed which the keyhole in the weld pool is unable to penetrate. In other words, by reducing the speed of welding at a given beam power, there is a certain material dependent depth of fusion zone, or seam thickness, which cannot be exceeded. This is illustrated in Fig. 62 by the upper limit of the series of curves.

4.6 Beam deflection

The electron beam is made up of negative charge carriers and can thus be deflected in different ways from its normal axis by electrical and magnetic fields. The following types of deflection are possible:

- Static deflection;
- Dynamic periodic beam deflection, also called beam oscillation;
- Dynamic aperiodic beam deflection.

If a direct current is passed through the coils of the deflection system (see section 2.7), then the electron beam will be deflected statically through a certain angle. Static beam deflection is required for making T-butt welds in the normal position. The required angle of deflection β is set by adjusting the direct current to the coils I_K (the deflection current). In general the maximum deflection angle β is $\pm 5°$ and is limited less by the increasing distortion at the point of impingement of the beam, than by the field of view of the optical viewing system.

With periodically varying alternating currents the electron beam can be deflected dynamically, free of inertia, in all directions, shapes and frequencies that the welding might require (Fig. 63). Beam oscillation at small amplitudes and high frequencies is a further effective and generally useful method of beneficially affecting the fusion and solidification processes with the aim of improving the quality of the weld produced. Beam oscillation causes the keyhole to increase in size, preventing the molten envelope from collapsing and reducing the resulting entrapment of voids in the root of partially penetrating welds. An oscillating electron beam in general increases the size of the fusion zone and allows gas porosity to rise and escape from the weld pool.

In addition beam oscillation can affect the dynamics of the molten zone, preventing the root of the weld from dropping through, and allowing the crown of the weld to solidify smoothly and without formation of sharp undercutting [60]. With suitable high amplitudes and low frequencies, beam oscillation will also permit axial circumferential seams to be welded, provided that with larger seam thicknesses and angles of beam deflection the danger of fusion defects is taken into account (Fig. 64). With this type of circumferential welding the radius error in the circular weaving pattern of the beam must be kept to $<0.25\%$.

A different type of dynamic beam deflection is required when welding with automatic path control and seam tracking systems, see Chapter 7. In these cases the deflection system is fed with an aperiodic alternating current by the beam control system, which permits the beam to follow a predetermined seam

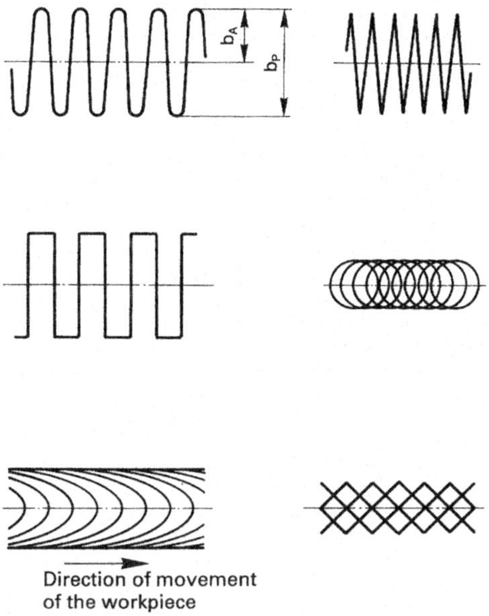

Fig. 63 Various ways of oscillating the electron beam. b_A = Amplitude of oscillating; b_P = Width of oscillation.

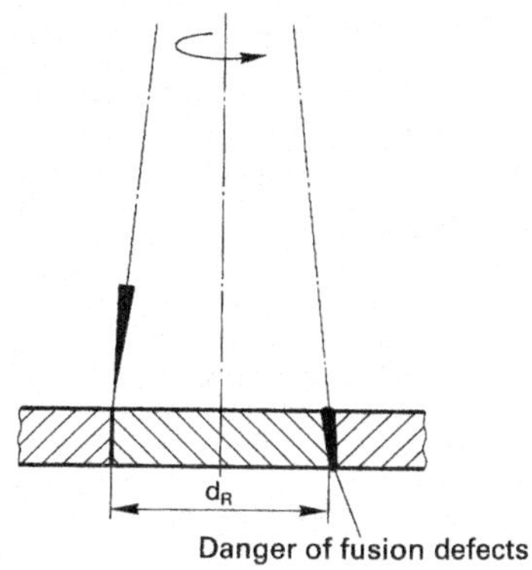

Fig. 64 Welding a circumferential weld seam with beam oscillation. The frequency of oscillation f_P and the diameter of the circumferential seam d_R determine the speed of welding v_S.

path which is angled in the X-Y direction, by the detection and correction of variations in the actual and required positions of the beam spot (for example trailing error correction).

Another particular type of beam deflection which must be mentioned is that which led to the tulip shaped seam technique for welding very thick workpieces. In this type of operation the electron beam is oscillated transversely whilst the beam current is modulated in various ways (Fig. 65).

With this distribution of beam power, the width of the seam can be controlled in the deeper regions of the fusion zone so that the root of the weld solidifies without problems, just as with thinner workpieces. A tulip shaped fusion zone with a broad cup and narrow stem is formed (Fig. 66 and 67). In this case, the term tulip seam technique has nothing to do with the similarly named method of weld seam preparation. At present there are no actual manufacturing processes using this technique. In trials using the tulip seam technique, trial specimens of 60 and 80 mm thickness have so far been welded. The angular shrinkage of the specimens did not nominally increase and on average was $<1°$. The hardness values measured in the boundary between the fusion zone and the heat affected zone in the lower part of the weld were some 10–20% above those at the top of the weld because of the higher rate of cooling (Material: 10 CrMo 9 10) [61].

Exactly which type of deflection, which frequency and which amplitude will produce the results mentioned can be determined only from a series of trial

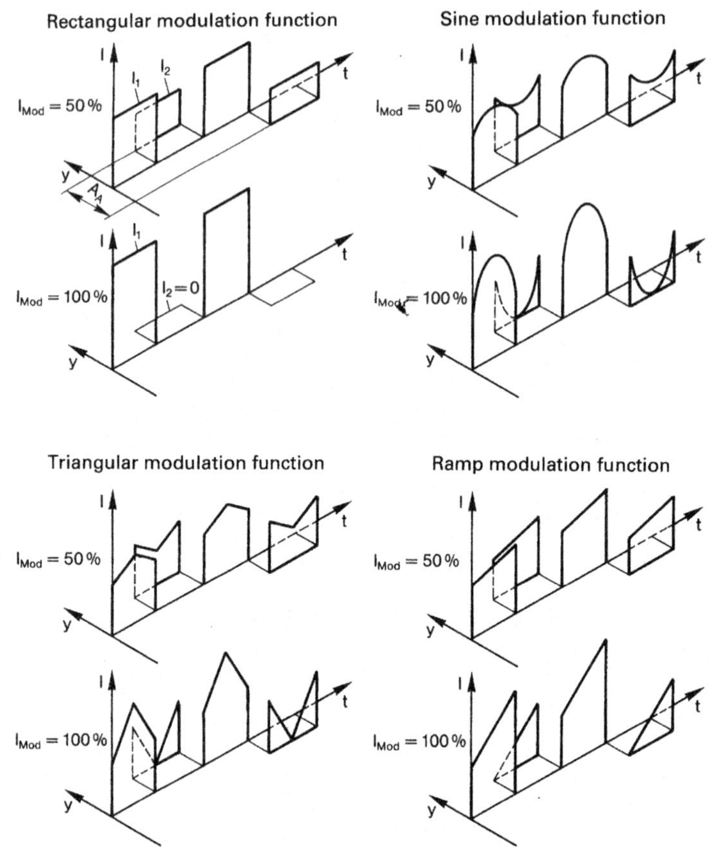

Fig. 65 Various ways of modulating the beam current when electron beam welding using the tulip seam technique [61].

welds. This is because of the numerous factors which influence the process such as the properties of the material in both the solid and liquid states (the viscosity of the melt), the speed of welding, the thickness of the material, the depth of the fusion zone and the power density and distribution, etc.

Thus in welding circumferential seams, the overlap of the preheat at the start of the weld and the gradual decrease in the depth of fusion in the final slope-out, all play an important role in determining the oscillation parameters and settings which must be used. From the literature [62–65] and from our own experiences, the following tips, which do not take into account the effects of the material to be welded, should give some idea of the types of effects which can be achieved:

- Longitudinal oscillation increases the depth of the fusion zone;
- Transverse oscillation, sine-, rectangular-, trapezoidal-, ellipsoidal- and X-shaped all stabilise the weld pool when welding with a root through

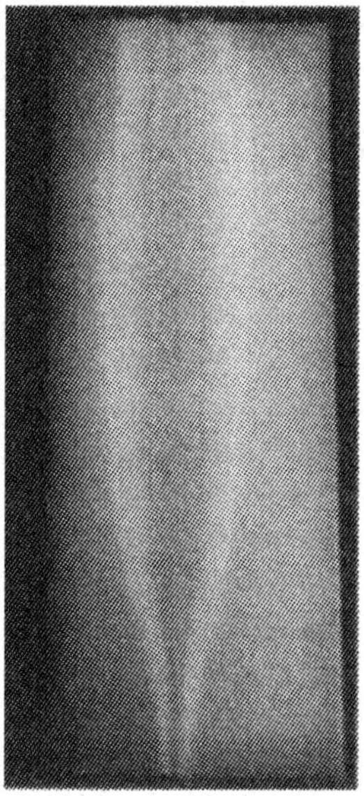

Fig. 66 An 80 mm thick weld in 10 CrMo 9 10 made using the tulip technique. Beam power P = 18 kW, welding speed v_S = 1.8 mm · s^{-1}. Shape of oscillation: Saw toothed.

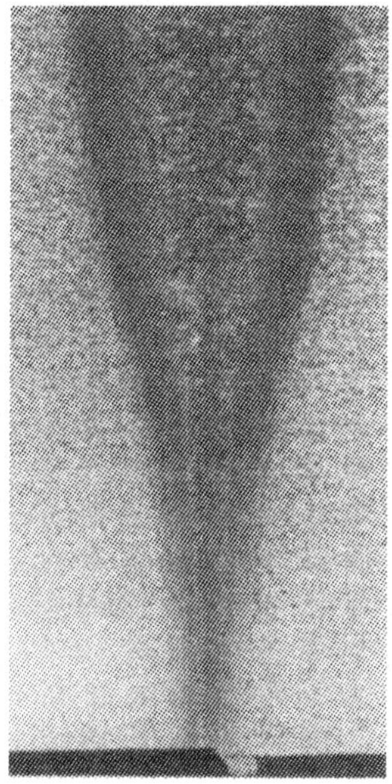

Fig. 67 Root area of the weld shown in Fig. 66. (3 : 1, Reproduced: 0.6 X).

the full thickness of the material, so that the range of optimum welding parameters is increased;

- Circular oscillation suppresses formation of root spiking (solidification cavities in the root);
- Parabolic oscillation smooths the crown of the weld and minimises undercutting;
- With deflection frequencies of $f < 25$ Hz, the electron beam moves the keyhole;
- With deflection frequencies of $f > 25$ Hz: the electron beam moves within the keyhole, thus expanding it;
- Deflection amplitudes investigated: b_p = 0.3 to 2 mm;
- With the exception of longitudinal- and X-oscillation, in all other cases the depth of the fusion zone is somewhat decreased when compared with similar welds made without oscillation. This effect can be compensated for by increasing the beam current and/or decreasing the speed of welding.

4.7 Beam pulsing

In the initial years of electron beam technology the maximum power of welding machines was limited to about 3 kW. There was a lack of experience and additional equipment to enable thick welds to be made at low welding speeds and to enable the beam to be oscillated in the many ways now possible. Thus, in spite of the deep welding effect thicknesses of up to only 10 and maybe 15 mm, depending upon the actual material, could be safely welded without defects. It was at this time that pulsed electron beam welding gained in importance. Today this technique for increasing the depth of the fusion zone is only of subsidiary importance.

With the aid of electrical circuitry in the high voltage generator, the beam current is repeatedly considerably increased and then switched off for short periods (Fig. 68). A pulsed beam current or pulsing beam power is thus produced having a certain defined frequency and high peak value, which by calculation is expressed as a mean value which does not exceed the maximum nominal beam power possible from the machine.

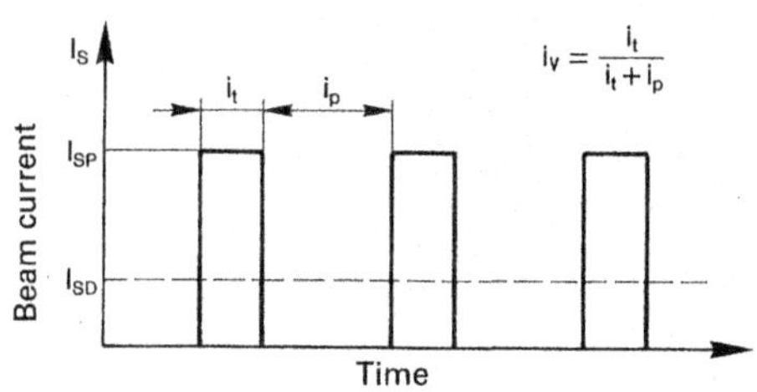

Fig. 68 The shape of a pulsed beam current: I_{SD} = continuous beam current, I_{SP} = pulsed beam current, i_t = pulse width, i_p = pause time, i_v = pulse ratio.

As a result of the high pulse current, the electron beam penetrates correspondingly deeply into the material, and through the close, overlapping nature of the individual keyholes so produced, as well as the diameter of the keyhole and the conduction of heat in the material, a compact and continuous fusion zone is produced in the direction of welding. The typical characteristics of pulsed weld seams are thus a large s/b ratio of up to 50 and an increase in depth of the fusion zone of about 50% compared with welds made with a continuous beam of the same power, but also a far more irregular and spattered weld seam. The increase of 50% mentioned indicates that the effect which pulsing the electron beam has on the depth of the fusion zone cannot simply be increased at will. On the one hand the fusion and heat conducting processes require a certain minimum time to ensure a weld containing no gaps. On the other hand, the pulsed current I_{SP} compared with the continuous current I_{SD} can only be increased within certain electron optical and instrumentation limits, depending upon the key ratio i_T.

These rules governing the use of beam pulsing also hold true for modern electron beam welding machines, although this technique is no longer used for increasing the depth of the fusion zone due to the much higher beam powers now commonly available. It is, however, used for production of very narrow fusion zones. A well known example of this is electron beam welding of the cooling channels in aluminium pistons (Fig. 69).

The weld must be made close to the cooling channel without damaging the wall of the channel by melting it. Removal of the crown of the weld is carried out by subsequently machining the outer surface of the piston. Another possible

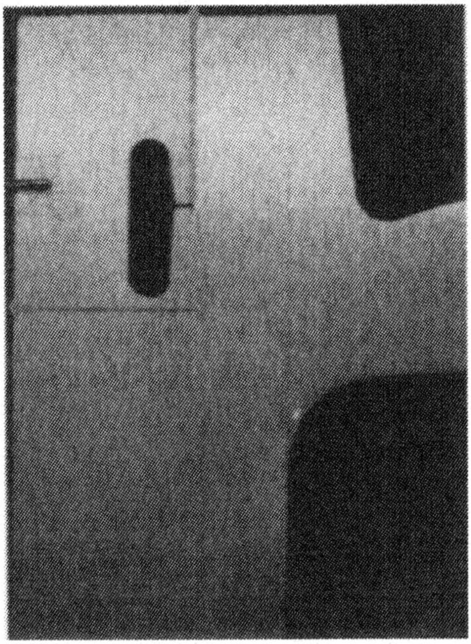

Fig. 69 Cross section through an aluminium piston welded with a pulsed electron beam (1 : 1, Reproduced: 0.5 X).

use for beam pulsing is in spot welding or 'stapling'. These processes are now carried out, using far greater ranges of settings than were initially possible, by the programmable computer control systems of modern electron beam welding machines.

4.8 Positional welding

In welding technology, the term positional welding is applied to all those geometrical positions of a workpiece which deviate from the 'normal' horizontal position (Fig. 70). When welding in position gravity acts on the weld pool in a different way from the horizontal position. Positional welding is often unavoidable due to the size of the workpiece and the position of the weld seams – for example in ship construction. In other cases it is used on purpose to improve control of a very large weld pool, for example in electroslag welding.

As far as electron beam welding is concerned, the structural reasons are the exception for positional welding. They require movable and rotatable electron guns built into very large working chambers in which the workpiece remains stationary.

As a rule, non-horizontal positions are chosen for electron beam welding of thicker workpieces. From $t = 50\,mm$ it becomes increasingly more difficult to keep the keyhole open, to prevent the melt from dropping through the workpiece and to achieve welds free of defects. With differentiated settings for various parameters such as defocusing, oscillating, etc, it becomes possible, depending upon the actual material, to bring the welding process under control

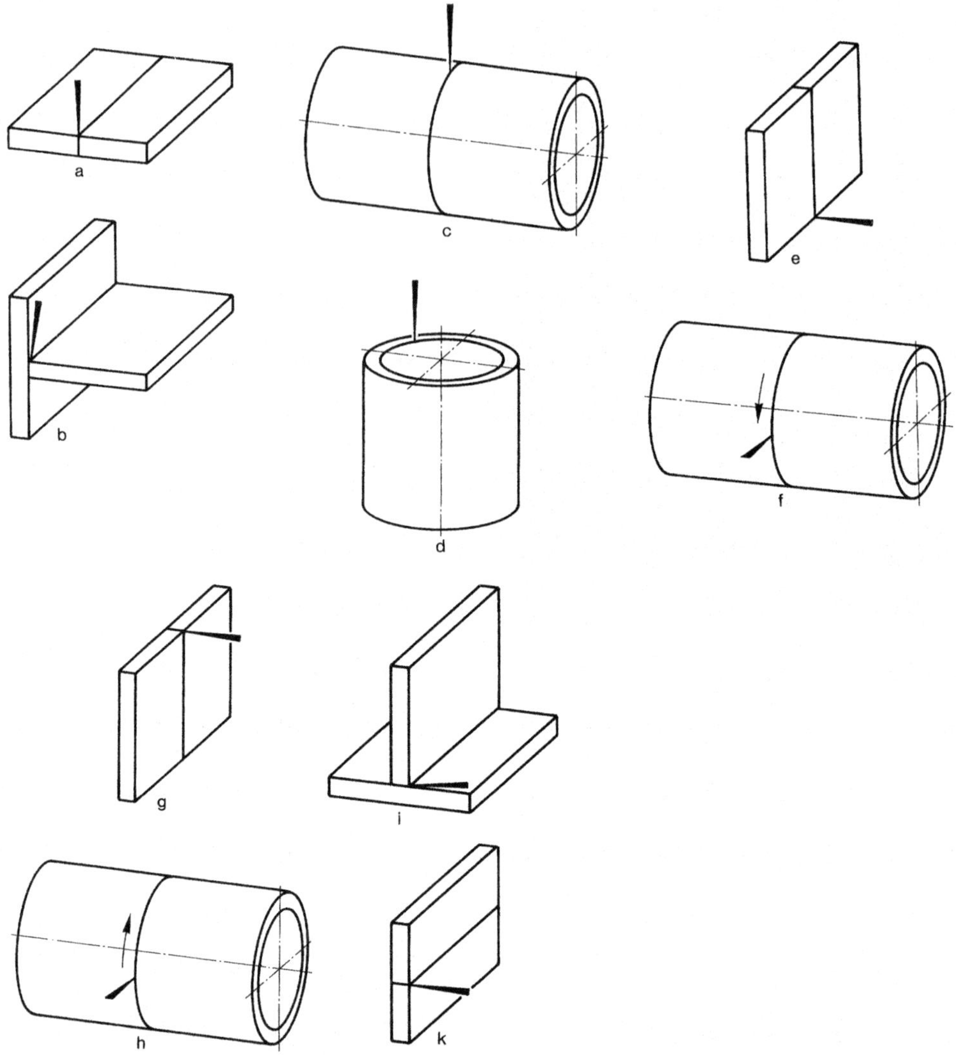

Fig. 70 The various positions used for electron beam welding: *a*) Horizontal (normal), symbol: w; *b*) Horizontal (inclined), symbol: h; *c*), *d*) Horizontal, symbol: w; *e*), *f*) Vertical-up, symbol: s *g*), *h*) Vertical down, symbol: f *i*), *k*) Transverse, symbol: q.

in the horizontal position even in material of thickness greater than t = 50 mm. For thicker welds there remains only the possibility of changing to a vertical-down (a 9 o'clock position in the case of circumferential welds) or transverse position.

Figure 71 shows how in a horizontal-vertical position the depth of the weld pool can be reduced and a stable keyhole achieved. Detailed investigations have confirmed these considerations so that today in manufacturing industry, nu-

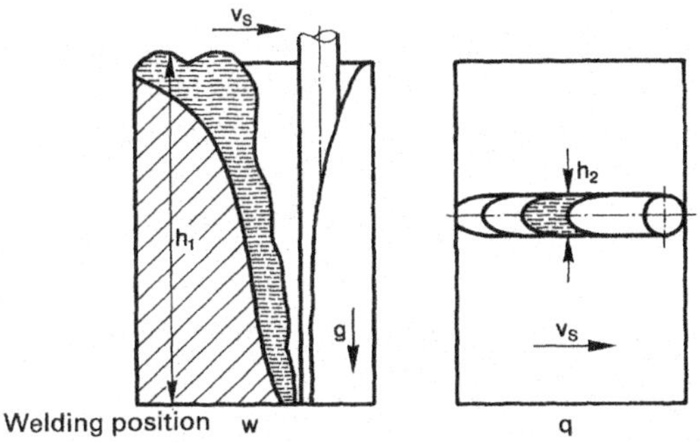

Fig. 71 Dimensions of the weld pool at different positions [61].

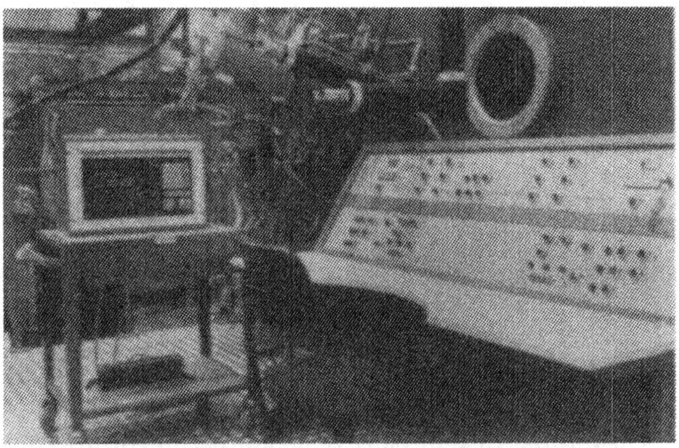

Fig. 72 An electron beam gun mounted on the vertical wall of the chamber for welding with an approximately horizontal beam.

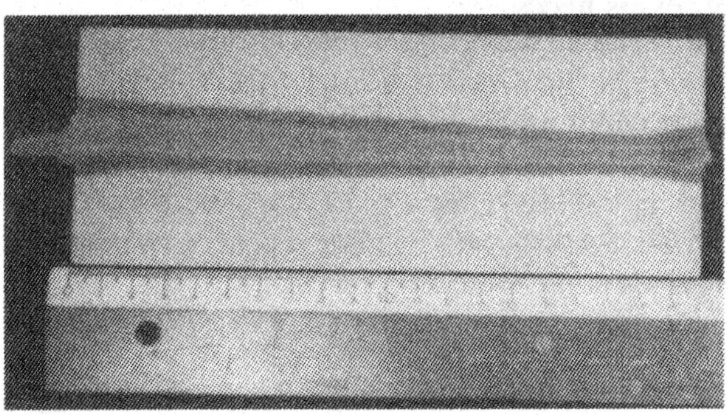

Fig. 73 Macroscopic cross section of an electron beam weld in 200 mm thick 10 CrMo 9 10. U_s = 160 kV, I_S = 400 mA, v_S = 1.1 mm · s^{-1}, welding position: q.

merous welds are successfully made in thick walled materials in the horizontal-vertical position (Fig. 72) [42, 66–68]. It has also proven advantageous to support the weld pool from the root face using a slotted backing strip. Figure 73, for example, shows a cross section through a 200 mm thick specimen plate of 10 CrMo 9 10 steel, exhibiting the lightly curved fusion zone typical of that produced when welding in the horizontal-vertical position.

4.9 Working pressure

In section 2.11 the reasons were given for a vacuum being an absolute necessity for generating and collimating an electron beam. On the path from the electron gun to the workpiece a vacuum cannot be dispensed with, but for other reasons an adequate pressure is required. The residual gases in the vicinity of the electron beam should:

- Not initiate any chemical reactions, such as oxidation at the point of welding and the heated areas of the workpiece;
- Not affect the power density in the electron beam, in other words they should not cause any scattering effects.

As numerous investigations have shown, to achieve these requirements a vacuum as low as $p_A = 5 \cdot 10^{-4}$ mbar is not necessary, since they can also be met at higher pressures.

If one chooses prevention of oxidation of the weld and the workpiece as the dominant criterion for deciding on a higher pressure for the working chamber, then as a comparable measure the purity of a normal shielding gas used in TIG welding of steel might be chosen, e.g. 99.99% pure welding argon. Regardless of which gases constitute the remaining 0.01% of impurities, a simple calculation shows that a vacuum of $p_A = 10^{-1}$ mbar would provide the same degree of purity.

With this vacuum steel can be electron beam welded under the same conditions of purity as when using the TIG process. For gas sensitive materials such as titanium, zirconium, etc, both methods of welding are required to meet higher degrees of purity. Thus in the case of electron beam welding a working pressure of $p_A = 10^{-4}$ mbar or less is necessary.

The second criterion, that of avoiding scattering effects, is based upon the fact that as the pressure increases the light electrons increasingly collide with the heavy gas molecules in the air and are thus deflected and scattered (Fig. 74). The direct result is that the diameter of the focal point is increased and thereby the power density is reduced. Figure 75, for example, shows that the depth of the fusion zone changes only slightly up to a pressure of $p_A = 10^{-1}$ mbar. Only as the pressure increases still further does the depth of the fusion zone begin to decrease significantly and more rapidly. This tendency is practically independent of both the material and the welding parameters. With otherwise identical conditions, however, the depth of the fusion zone begins to decrease in all cases at pressures greater than $p_A = 10^{-2}$ mbar, so that from the point of view of welding technology, this value must be seen as the absolute maximum which can be tolerated in the working chamber – or in other words – as the maximum permissible pressure for electron beam welding of normal materials (steels, etc).

In order, practically, to carry out welding at these pressures, the electron beam gun must be equipped with a pressure stage to maintain the pressure difference between the electron beam gun and the working chamber (see Fig. 30). For electron beam welding, which requires the very highest reproducibility

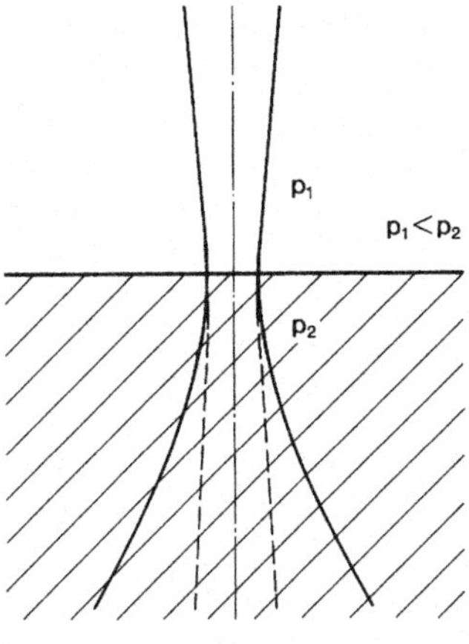

Fig. 74 Scattering of an electron beam on entering a medium of higher pressure ($P_2 > 10^{-2}$ mbar).

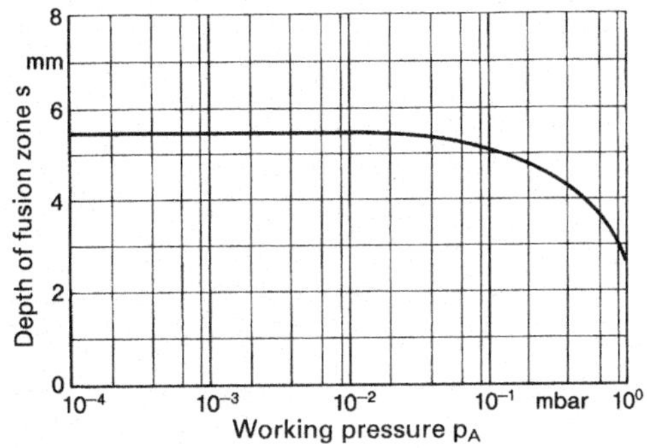

Fig. 75 Relationship between the depth of the fusion zone s and the working pressure p_A.

of fusion zone shape, it is also useful to incorporate other pieces of instrumentation to ensure that welding is always carried out at the same working pressure.

Very early on, because of the vacuum required to produce an electron beam, the question arose as to whether the need to evacuate the working chamber also could somehow be obviated. Since then it has been demonstrated many times just how and at what equipment cost it is possible to carry out electron beam welding in air at normal atmospheric pressure. The first requirement is a number of pressure stages equipped with powerful vacuum pumps to lead the electron beam out of the gun and towards the workpiece which must be positioned closely beneath it. As will be understood from the explanations given so far, the electron beam is scattered quite considerably in air such that even with high beam powers only relatively small weld thicknesses can be achieved. It is easy to understand why, for economic reasons, electron beam welding in air has not been adopted in Europe. The pressure stages and additional vacuum pumps required constitute a considerable capital investment, even with the saving of not requiring a vacuum working chamber.

Radiation shielding is also needed, the weld and its surroundings must be protected from oxidation by additional gas shielding, and even in the case of small thicknesses relatively high beam powers are necessary.

4.10 Optimisation of parameters

Determining the optimum parameters before starting a new welding operation can involve a very wide range of considerations. Not only must a weld of suitable thickness be produced to fulfill the functions and loadings expected of the component, but it must also be free of both external and internal defects. The settings required to achieve this are thus best determined in two stages if previous parameters for similar types of welding are not available. In general, as has already been mentioned in section 4.2, the accelerating voltage is first set to U_{Bmax} and thereafter kept constant. It is also recommended that initially a lens current I_L is chosen which produces a slight under-focusing. Apart from this the beam current I_s, depending on the as yet still unknown welding speed v_s, must also be determined. Each of these parameters has a very complicated effect on the welding process and thus on the weld itself. For example, in some cases the optimum beam current between being 'under' and 'over' welded may lie within very close limits.

As a result of this, other manufacturing tolerances and machine settings such as normal thickness tolerance, variations in gap width, deviations in welding speed, etc, may have a strongly adverse effect on the reproducibility and quality of the weld. On the one hand, the tolerance on the beam current setting can be increased by increasing the speed of welding or beam oscillation. On the other hand, surprisingly, a further increase in beam current at the same speed of welding can assist in closing a root previously subject to spatter and in elimination of defects. The success of the particular action taken depends to a great extent upon the material, the thickness of the weld and the speed of welding, etc. In other words, in the absence of any previous similar experience, the solution can only be determined by carrying out a series of welding trials.

It is often asked whether it is possible to determine the parameters for welding by calculation rather than from welding trials. Published mathematical formulae [44, 52, 53] are, however, unable to offer any assistance for two reasons. It is not the intention of these calculations to predict the required settings for a given welding operation. The purpose of such calculations is to establish whether the assumptions made regarding the penetrating behaviour of the electron beam or the energy transformation and physical processes taking place in the keyhole and molten envelope can be confirmed, and whether they agree with other known rules and relationships.

This is achieved by comparing the calculated results with those obtained experimentally. In doing this only one of many different welding operations will be chosen, often the production of a partially penetrating weld, with the only criterion considered being that of the depth of the fusion zone.

Because all such mathematical calculations involve a certain degree of simplification of the processes taking place, they are ultimately only able to provide the practical welder with a rough estimate of values. Effects such as the geometry of the component, the shape of the weld seam, the de-focusing and the focal distance are only partially taken into account. Thus the term 'beam current' needs to be defined more accurately to avoid misunderstandings and errors when settings are compared. With few exceptions the term beam current refers to a value read from the operating console of the welding machine, which is measured using calibrated resistances in the high voltage circuit. These values give an accurate enough measurement of the emission current at the cathode, but not always of the current impinging upon the workpiece, which is of greatest significance to the welding process. The impinging current, as can be demonstrated by actual measurement, see section 11.3.3, is somewhat less than the emission current.

	Test report	EB-process	Job-No.	Sheet

Description of testpiece	Shape and position of weld seam, fusion path or hardness trace

Test parameters

Machine and generator	Lens: single ○ double ○	CNC: Yes ○ No ○	CNC data see sheet

Vers. Nr.	U_B kV	I_s mA	I_L mA	V_s mm/s	A_A mm	*Beam oscillating Shape	f_P Hz	b_P mm	Run-in s	Rotation s	Overlap s	Run-out s	

Comments on welding process

Examination of the specimens

Test No.	Top-/Front- pass	Bottom-/Rear- pass	Externally visible cracks		Rad, see sheet	U/S, see sheet	Met, see sheet	Hardness, see sheet

Comments on results of examination

Fig. 76 Example of a test protocol for determining the welding parameters required.

EB-Welding documentation | File No.

Date ....

Details of Joint (sketch)

Weld thickess t = ... mm
Depth of fusion zone s = ... mm
Weld gap b_s = ... mm

Welding position:
(horiz; axial, vertical)

h ○ a ○ v ○

other

Customer: ....

Weld direction

Workpiece ....

8 0 0

8 5 0

Report No.

Straight ○ A ○ B ○ C ○
other ....

Chamber vacuum: P_A = ... mbar

Filler yes ○ no ○

Insert ○ Wire ○

Thickness .... mm Ø mm

Parent material ....

Material spec. ....

Chemical composition ....

Backing strip yes ○ no ○

Filler material ....

Welding machine ....

Material spec. ....

Preheat yes ○ no ○

Chemical composition ....

type

temp. .... °C

General findings, microstructural examination (photograph).

Accelerating voltage U_B = ... kV
Beam current I_S = ... mA
Beam pulsing yes ○ no ○
Pulse frequency i_f = ... Hz
Working distance A_A = ... mm
Lens current I_L = ... mA
Focus relative to
surface of workpiece u = ... mm
Focal distance A_F .... mm
Beam weaving yes ○ no ○
Function-type:
Width of weaving b_p = ... mm
Frequency of weaving f_p = ... Hz
Welding speed V_S = ... mm/s

Tack weld yes ○ no ○
Beam current I_S ....mA
Welding speed V_S .... mm/s

Beam run-in and run-out: t_E ...... s
t_A ...... s

Engineer

Fig. 77 Example of a weld procedure sheet for the electron beam welding of a component.

The magnitude of beam current losses is dependent upon many different factors, for example machine dependent effects, such that in spite of the same beam current being indicated, the impinging current delivered by welding machines produced by different manufacturers may in fact differ. Thus it can be seen that the ability to transfer calculated and even experimentally determined beam current values to all types of electron beam welding machine is extremely limited. In contrast, the reproducibility of such data, once set, is one of the great advantages of the process which results in welds both of high quality and uniformity.

The second stage in optimising the welding parameters is to eliminate any external (undercutting) and internal (pores) defects that might be present. Initially this may be achieved by beam oscillating, by altering the main beam current parameter or the lens current, or by changing the speed of welding. This is done in such a way as to ensure that the crown of the weld solidifies without any undercutting and, by increasing the size of the weld pool, by allowing sufficient time for pores either to rise to the surface or to close. The complex interdependence of the beam parameters and the properties of the material in its different conditions may initially appear somewhat bewildering. Actual practice shows, however, that with increasing experience the steps required to produce a perfect weld rapidly diminish.

4.11 Welding protocols and documentation

Successful and economic electron beam welding also requires the use of protocols, test reports and weld procedures. The structure and relationship of these types of form depend equally upon their intended purpose, such as process development, strength determination, preparation for manufacture etc, as on the significance of the respective parameters in terms of the aim of the operation. Figure 76 shows an example of such a test report, which is intended to provide a starting point from which a more suitable form can be developed. The settings used in the trials are then extracted from the protocol sheets and reports and released to the final manufacturing operation as a weld procedure.

The weld procedure sheet shown in Fig. 77, also intended only as an example, is particularly important in the case of electron beam welds in components which require third party acceptance. In these cases such weld procedures form part of the acceptance documentation and may not be altered or amended at will. In addition, a complete set of carefully prepared protocols and reports is a valuable help in design and preparation of new welding processes and operations.

5 The weldability of metallic materials

5.1 General

One of the main requirements for reliable and economic welding is a knowledge of the properties of the material and how they are affected by the thermal processes taking place. In most cases, whether or not a material is suited to welding and to what extent its characteristic mechanical properties might be changed depend upon numerous factors such that based upon only general information an overall appraisal is not possible. Even in the case of specific groups of materials, it is as difficult to give a clear yes or no answer to the question of weldability as it is to quantify the changes which might occur in the dynamic material properties on welding. With an unknown material, depending on the function and value of the component, it must be decided whether sufficient information is available on its metallurgical behaviour during electron beam welding, or whether welding trials are need. These might be carried out on the actual component, or on a similar type of specimen, in order to obtain the necessary information regarding weldability and strength.

The following sections thus describe the general behaviour of some chosen metals with respect to the processes involved in electron beam welding, with reference to investigations which deal in detail with each of the individual themes.

At this point it should be mentioned that the weldability of a given component can only be judged to a limited extent from the weldability of its material. According to the terms referred to in DIN 8528 Part 1, the weldability of a component is determined equally by its suitability for welding and the welding processes to be employed during its manufacture.

5.2 Process related effects

5.2.1 The liquid and vapour phases

Melting and vaporisation of a material at the start of the welding process is measured in milliseconds [37]. This rapid heating process has a mostly indirect effect on the behaviour of the material. The time for which the volume of material concerned is exposed to such high temperatures is very short. Pyrometric measurements at the point of impingement of the electron beam have revealed temperatures of about 2800 K, which are thus still below that of vaporisation [28]. Similar results were obtained from investigations on aluminium based materials on which surface temperatures of some 1300–

2200 K were measured [69]. In steels, temperatures of about 3800 K have been estimated at the bottom of the keyhole [9]. In spite of the short exposure times to this range of temperatures, as a rule, and depending upon the chemical composition of the metal, the gaseous elements present have sufficient time to escape before solidification begins.

On the other hand, in the case of materials which are susceptible to formation of porosity, this degassing phase must be extended by beam de-focusing, oscillation or repeated welding.

5.2.2 The effect of the vacuum

Electron beam welding, with few exceptions, is carried out in a vacuum. The vacuum intensifies the degassing process mentioned, which occurs during the liquid and vapour phases. In general this improves the properties of the material concerned, which is why vacuum degassing of certain metals is widely practised in industry. The electron beam welding process is, however, not the most suitable means of improving the fusion zone of a metal by vacuum refining. On the one hand, in the case of certain special steel, nickel and cobalt alloys, vacuum melting is a requirement in their manufacture to ensure porosity and crack free electron beam welding. In these cases, the vacuum effect occurring during welding is insufficient to permit adequate degassing during the short liquid and vapour phases. On the other hand, certain aluminium and copper materials contain high vapour pressure alloying additions such as magnesium and zinc. These escape in such large quantities during electron beam welding that the previously mentioned beam de-focussing is, for technical reasons, insufficient to prevent the process of degassing.

A particular aspect of the vacuum effect to which relatively little attention is paid must also be mentioned. As is generally known from welding with tungsten electrodes under a shielding gas, addition of 1–3% oxygen to the argon significantly reduces the surface tension of the weld pool. This effect is also noticed in electron beam welding, where the complete absence of oxygen causes the weld pool to become very viscous and, together with violent motion within the weld pool and high rates of cooling, causes undercutting and very distinct scalloping of the crown of the weld on solidification (Fig. 78).

Fig. 78 Crown of an electron beam weld showing scalloping (8 : 1, Reproduced: 0.6 X).

5.2.3 Solidification

Extremely rapid and localised heating by the electron beam leaves the material adjacent to the weld almost unaffected thermally. As a result of its low temperature, this material is thus able to conduct away the heat in the fusion zone in a time much shorter than in any autogenous or arc welding processes. Maximum published [70] values for the rates of cooling are of the order of 10^3–$10^5\,K \cdot s^{-1}$. The result is that the melt, including the condensing component, solidifies to form an extremely fine grained structure which, in spite of the high starting temperatures, shows no evidence of overheating which indicates the presence of embrittlement in high alloy steels.

The extremely rapid solidification can act very detrimentally on the closing of the keyhole and degassing of the melt. In particular, as has been mentioned in section 3.3, when welding seams which do not penetrate through the full thickness of the material, unstable processes take place at the base of the keyhole. These can trap voids in the melt which, because of lack of time and energy, cannot be reopened by the electron beam. If at this moment there is also insufficient molten envelope available to follow the shrinking void, then as the material moves because of the shrinkage occurring on solidification, irregular and often sharp edged cavities are formed. As far as the strength of the material is concerned, such cavities represent dangerous internal notches and stress raisers (Fig. 79).

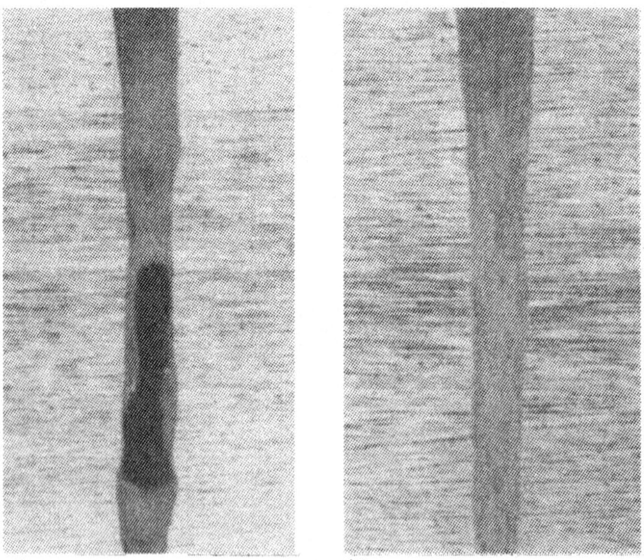

a) b)

Fig. 79 Cross sections through welds in high alloy steels (8.5 : 1): *a)* With solidification void; *b)* Free of voids because of beam oscillation.

In comparison, pores need not be judged quite so critically because of their mostly rounded or elliptical cross sectional shape. These are formed in the melt either as a result of metallurgical gaseous reactions, for example when welding high vapour pressure alloying components, or during cooling as a result of the

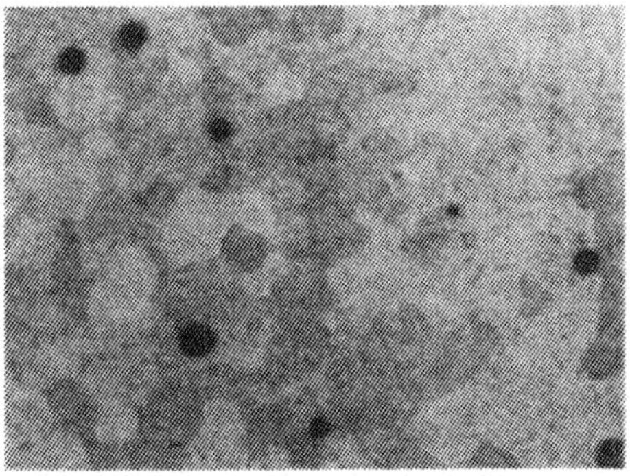

Fig. 80 Microsection of a weld in TiAl 6 V 4 showing porosity (200:1, Reproduced: 0.6 X).

decreasing solubility of gases. Because of the rapid solidification of the melt, gases cannot escape nor can the pores close up (Fig. 80).

As has been mentioned many times, there is a whole series of beam parameters, such as beam de-focusing, beam deflection and repeated welding, etc, which can be used largely to suppress the formation of cavities and pores during solidification.

Macrosections taken parallel to the surface of the material, depending upon the particular material group, exhibit more or less distinctive solidification rings which indicate the direction of cooling of the melt (Fig. 81*a*). Cooling begins at the boundary with the parent metal and finishes at the centre of the weld, whilst simultaneously the direction of solidification changes. Detailed investigations have shown that the rings become more separated as the beam current increases, and closer together as the speed of welding increases. There is no primary relationship between the solidification rings and any mechanical vibration in the workpiece or irregularity in the heat input during welding. The

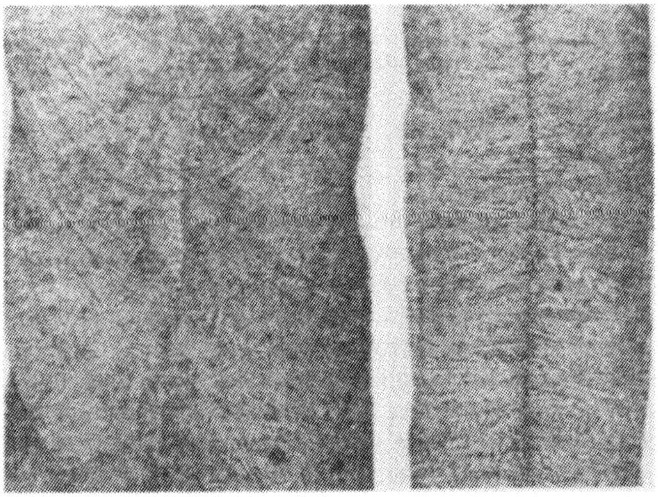

a) b)

Fig. 81 Horizontal sections through welds in X 2 NiCoMo 18 9 5 (80 : 1, Reproduced: 0.6 X): *a*) Solidification rings formed at $v_S = 10\,mm \cdot s^{-1}$; *b*) Solidification plane formed at $v_S = 30\,mm \cdot s^{-1}$.

solidification rings are a result of impurities and alloying elements which cause undercooling, or in other words locally delay solidification of the melt, the fusion heat from which produces zones of differing dendritic structure. The rings are thus the result of periodically varying rates of solidification [72]. Small areas enriched with impurities and alloying elements, which occur along the solidification rings, are insignificant.

If the enriched areas become more apparent, or join together, for example, to form a continuous line at the centre of the weld (Fig. 81*b*), then this may adversely affect both the strength and corrosion resistance of the material (Fig. 82). This rather detrimental type of solidification can be suppressed by an additional periodic movement of the melt (beam oscillation).

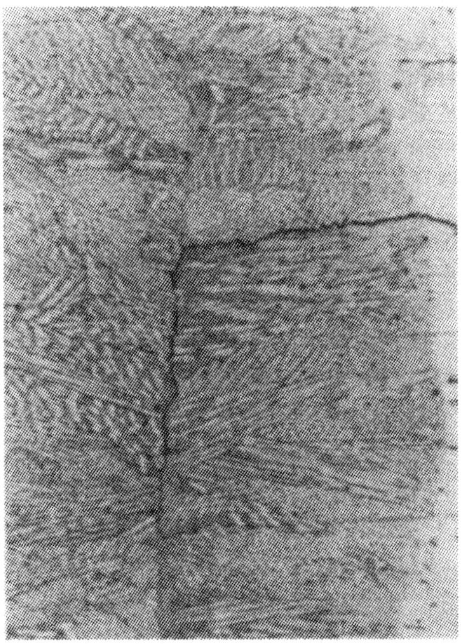

Fig. 82 Horizontal section through a weld in an austenitic steel (50 : 1, Reproduced: 0.5 X).

5.2.4 Heat affected zones

The regions of the material which remain solid during welding are subject to a heating and cooling cycle the peak temperatures and temperature gradients of which are quite different from those in the fusion zone. The maximum temperature reached in the solid material is lower the greater the distance from the fusion zone (Fig. 83 and 84). The microstructural changes occurring in the solid material, as a result of the extremely rapid temperature cycle, also mean that no coarsening of the structure takes place. With highly alloyed steels this can be extremely beneficial to the corrosion resistant properties of the material [73]. On further cooling, the temperature range from 800 to 500 °C is reached and this is of particular importance as far as the hardness of structural steels is

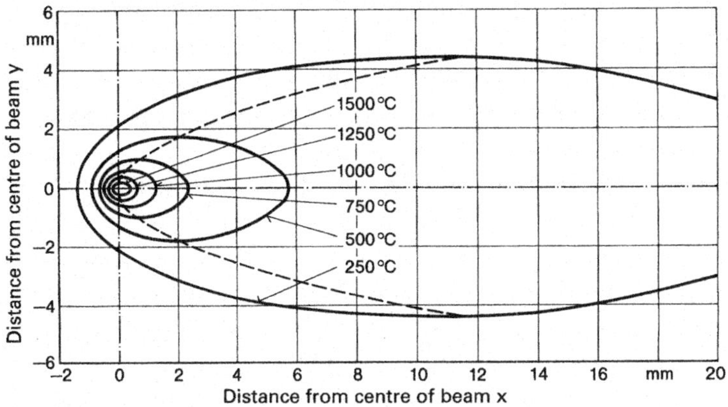

Fig. 83 Quasi-stationary temperature field when electron beam welding St 42; $t = 8\,mm$, $v_S = 6.4\,mm \cdot s^{-1}$ [71].

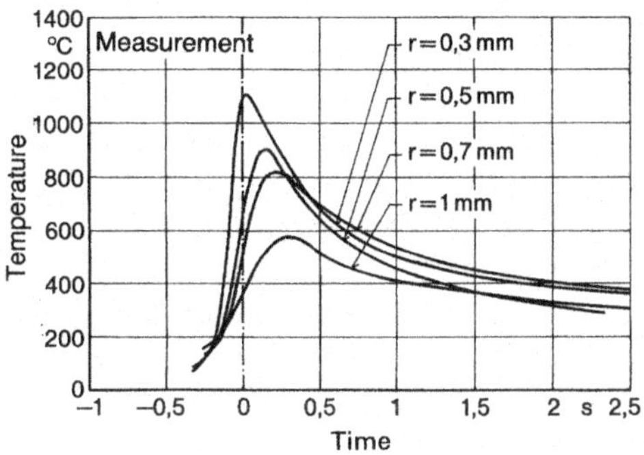

Fig. 84 Time – temperature curves at various distances from the centre of a weld in St 52 [71]. $t = 8\,mm$, $v_S = 6.4\,mm \cdot s^{-1}$.

concerned. Next come the temperature ranges which affect strengthening by precipitation hardening.

In all evaluations of the fusion and heat affected zones it must be remembered that the microstructural and strength changes which occur are limited to a very narrow band of material in comparison with other welding processes. The cast microstructure of a fusion zone and the hardness peaks or troughs in the weld and heat affected zones only slightly affect the strength and elongation of thc joint, especially under static loading conditions. If the weld is loaded dynamically, however, then as is the case with other methods of welding and also with heat treatments in general, a loss in toughness of the cast microstructure of the fusion zone (a metallurgical notch effect) is to be expected.

5.2.5 Internal welding stresses

During the heating and cooling which occur on welding, stresses are set up. Internal welding stresses are those which remain in the welded region of the component after cooling to room temperature. Weld stresses or internal weld stresses can cause formation of different types of cracks (hot cracks, shrinkage cracks, etc), and are in all cases the main cause of component distortion. A knowledge of the cause of these types of stress will thus help both to prevent weld defects and to maintain the tolerance of finished components on welding [74].

The magnitude of internal welding stresses depends not only upon the welding process and thus the welding parameters and shape of the weld, but also upon:

- structural factors such as the geometry and stiffness of the component;
- material factors such as the coefficient of thermal expansion, the metallurgical behaviour, the yield point, etc.

Welding stresses are a result of localised heating and prevention of the processes of expansion and shrinkage which are typical of any fusion welding process. Figure 85 shows how these processes act. In electron beam welding, the weld seam causes thermal expansion of the solid material of the parent metal. This causes the adjacent colder parent material to be compressed plastically because, as a result of the geometry of the component, these stresses can only be released to limited extent (Fig. 85*b*). After welding, the heated parent material cools and contracts whilst the compressed parent material which was not heated during the welding process remains compressed (Fig. 85*c*). If the component is now able to shrink unhindered then it will shorten, which is the simplest form of welding distortion. Whilst Fig. 85 illustrates the process of transverse shrinkage, this can also occur to a far greater extent along the weld and to a lesser extent through the thickness of the weld.

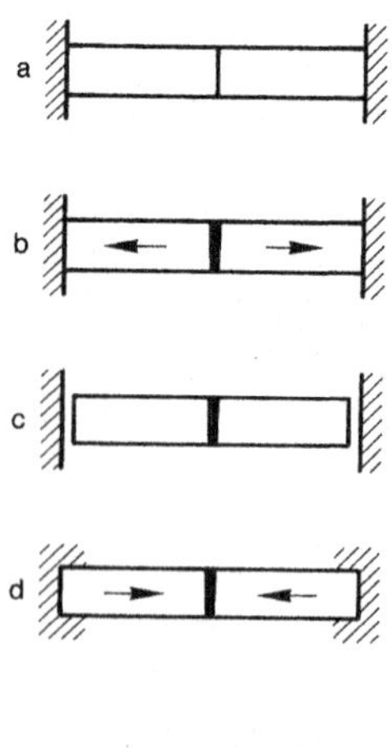

Fig. 85 The formation of transverse shrinkage and internal shrinkage stresses on welding: *a*) Before welding; *b*) Expansion during welding; *c*) Unrestrained shrinkage; *d*) Shrinkage prevented.

If the shrinkage is also prevented by the geometry of the component (Fig. 85*d*), internal shrinkage stresses will be set up which also form part of the overall pattern of internal welding stresses (see section 6.12).

A relationship exists between the internal shrinkage stresses and the resulting distortion which is often not immediately apparent. Small distortions occur when shrinkage is prevented or – with respect to electron beam welding – when the internal stresses are low. Low internal shrinkage stresses are set up whenever shrinkage is unhindered or, for example, whenever the weld volume is small. A component exhibiting low distortion can thus nevertheless contain either very high internal shrinkage stresses (high component stiffness prevents shrinkage), or low shrinkage stresses (narrow electron beam welded joints). In contrast, a component which is very distorted after welding is always characterised by low internal shrinkage stresses.

The actual relationships which arise, as in so many cases, are considerably more complex. It must be remembered that the weld does not cool uniformly as has just been assumed, but cools more rapidly at its surface than internally. As a result irregular internal stresses are set up, so called internal quenching stresses, which on locally exceeding the yield point of the material cause plastic deformation processes to take place. Furthermore, in certain materials phase transformations involving changes in volume may also play a role. This type of transformation stress occurs in steels whenever martensitic or intermediate transformations occur at sufficiently low temperatures [75].

All three of these processes can be superimposed upon one another, leading to very different local values of internal stress. In the very simple weld geometry of electron beam welding these stresses can be easily calculated, whilst in other welding processes they can only be determined experimentally

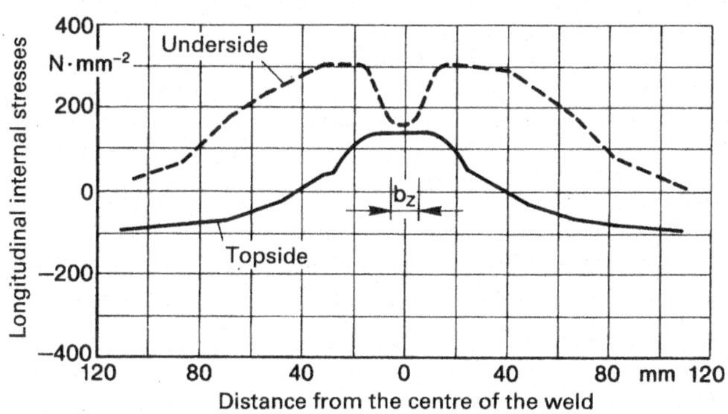

Fig. 86 Longitudinal internal stresses in a submerged-arc weld. Material: 22 MnMoNi 5 5. Specimen dimensions 220 × 250 mm, t = 75 mm, DY-seam [75].

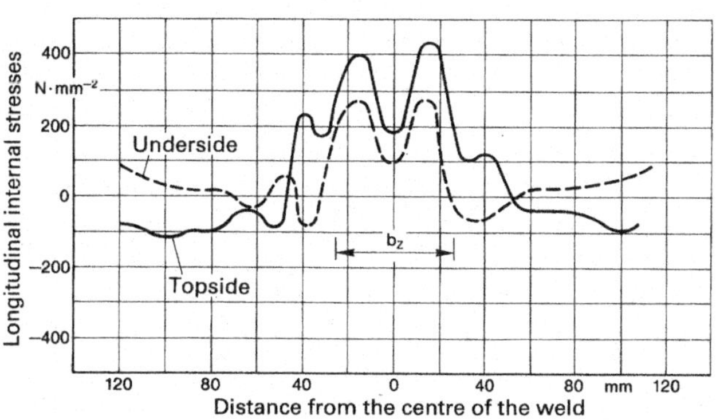

Fig. 87 As in Fig. 86 but for an electron beam I butt weld.

using different methods of measurement. Figures 86 and 87 show results of internal stress measurements made on electron beam welded 20 MnMoNi 5 5 plate material compared with the same material welded using the submerged-arc process. Particularly apparent is the strong dependence of the internal stresses on the distance from the weld and on the thickness of workpiece. Especially in thicker workpieces, internal stresses can be set up which approach the yield point of the material ($R_{p0.2} = 460\ Nmm^{-2}$ in the annealed condition), and which are caused by prevention of shrinkage and by the high rate of cooling. Prevention of shrinkage also occurs in both circumferential welds and rigidly clamped workpieces, which explains the increased tendency of some such materials to hot cracking.

A number of test methods have been developed with which the tendency of a material to hot cracking on electron beam welding can be tested [76, 77]. In quenched and tempered steels, however, internal transformation stresses are predominant. If electron beam welds are compared with submerged-arc welds in the same material it becomes apparent that narrow fusion zones and high welding speeds minimise internal stresses [75]. Within the scope of these investigations electron beam welds exhibited overall lower longitudinal and transverse internal stresses together with smaller internal stress differences, each of these factors being responsible for the minimal distortion observed.

5.3 Classifying weldability

What is meant by the term weldability? Most obviously it is a question of the behaviour of a material on welding. In an easily weldable material one expects those properties which will promote 'defect free' welding of the material, that is freedom from cracks, pores, etc. The material alone cannot guarantee defect free welding, however, because there is a range of other causes of welding defects which are not material based. Weldability is apparently the interaction of a range of different material properties which result from the chemical composition of the material, its method of production (unkilled, killed, vacuum smelted), the deformation processes undergone and its heat treatment, etc.

These factors all go together to determine the particular characteristics and mechanical properties of the welded joint produced. Whether or not a material is suitable for welding can initially be determined by testing the characteristic properties and mechanical strength values after welding. If the test results prove unsatisfactory with respect to the actual requirements, then technical changes can be made to the weld (e.g. widening the seam) and/or mechanical means adopted to improve the weld (e.g. use of a function generator to produce beam deflection or welding with additives or filler materials). From this it will be clear that weldability is essentially a material property but that in electron beam welding both the method of welding and the equipment required should also be considered.

For the purposes of the following discussion the weldabilities of various metallic materials are classified in accordance with DVS-Merkblatt 3204 as follows:

A – Weldable using normal practice.
Normal practices are: De-focusing, beam oscillating, beam offset, use of stress relieving slots adjacent to the weld, etc.

B – Weldable using particular techniques.
Particular techniques are: Pre- and post-weld heating, welding using additives or filler materials, etc.

C – Of limited weldability.
Welds in these materials are not evaluated in DIN 8563 Part 11 and are only for use in limited applications.

Table 2 lists details of the electron beam weldability of a number of metallic materials.

Table 2 The relative electron beam weldability of metals (from information sheet DVS 3204)

Material No.	Identification	Weldability	Depth of weld examined, mm
1. Steel			
1.0038	R St 37-2	A	20
1.0060	St 60-2	B	10
1.0401	C 15	A	10
1.0460	C 22.8	B	50
1.0526	C 45	B	5
1.0570	St 52-3	A	20
1.1191	Ck 45	B	10
1.2241	51 CrV 4	A	2
1.4006	X 10 Cr 13	A	20
1.4057	X 22 CrNi 17	A	10
1.4303	X 5 CrNi 18 12	A	5
1.4404	X 2 CrNiMo 17 13 2	B	50
1.4541	X 6 CrNiTi 18 10	A	50
1.4580	X CrNiMoNb 17 12 2	B	50
1.4923	G-X 14 CrCoMo 12 10	A	10
1.4980	X 5 NiCrTi 26 15	A	5
1.5920	18 CrNi 8	A	10
1.6358	X 2 NiCoMo 18 9 5	A	10
1.7131	16 MnCr 5	A	10
1.7335	13 CrMo 4 4	A	50
1.8902	StE 420	A	20
2. Copper materials			
2.0040	OF-Cu[1)]	A	25[2)]
2.0070	SE-Cu	B	25[2)]
2.0090	SF-Cu	B	25[2)]
2.0916	AlBz 5	A	18
2.1030	SnBz 8	A	5

Table 2 (Cont.)

Material No.	Identification	Weldability	Depth of weld examined, mm
3. Nickel materials			
2.4375	NiCu 30 Al	A	5
2.4858	NiCr 21 Mo	A	30
2.4951	NiCr 20 Ti	A	10
2.4969	NiCr 20 Co 18 Ti	B	10
4. Aluminium materials			
3.0255	Al 99.5	A	5
3.0515	AlMn 1	A	5
3.1355	AlCuMg 2	A	10
3.2315	AlMgSi 1	A	12
3.2581	G-AlSi 12	A	40
3.3535	AlMg 3	B	5
3.3547	AlMg 4.5 Mn	A	1.5
3.3555	AlMg 5	A	10
5. Titanium materials			
3.7035	Ti 2	A	50
3.7165	TiAl6 V 4	A	50
3.7175	TiAl6 V 6 Sn 2	B	25
3.7255	Ti 3 Pd	A	50
6. Material combinations			
1.0401 with	C 15 with	B	7
1.1173	Ck 34		
1.0570 with	St 52-3 with	A	6
1.7321	20 MoCr 4		
1.2003 with	75 Cr 1 with	A	2.5
1.3342	SC 6-5-2		
1.4006 with	X 10 Cr 13 with	A	20
1.7335	13 CrMo 4 4		
1.4303 with	X 5 CrNi 18 12 with	A	5
2.4668	NiCr 19 NbMo		
1.5919 with	15 CrNi 6 with	A	2
1.7147	20 MnCr 5		
7. Special metals and metal combinations			
	Mo (cast)	B	3.5
	Mo with Ta	B	1.6
	Mo 0.5 Ti	A	2.5
	TZM (Mo-alloy)	A	8
	Nb	A	1
	Nb with Ta	A	0.2
	Nb with Ti	A	3
	Nb 1 Zr	A	3
	Ta	A	3
	T	B	3.8
	Zircalloy-2	A	15

[1] P and O_2 contents <5 ppm
[2] No heat treatment necessary with T >200°C after welding.

5.4 The weldability of individual groups of materials

5.4.1 Irons and steels

Most iron and steel materials are very suitable for welding, and so in the following section only certain particular points are dealt with.

The restricted number of unkilled steels still in use are unsuitable for electron beam welding. As a rule they lack the additives required to suppress the gaseous reactions which otherwise take place in the weld pool. Just as unsuitable are the free machining steels due to their high sulphur contents. Case hardening and nitriding steels can be successfully electron beam welded prior to surface treatment or after machining away the surface treated regions in the area of the weld, see section 6.2. As the carbon content increases the tendency to hardening also increases. As a result of the high rates of cooling experienced during electron beam welding hardening becomes significant at about 0.2% carbon content, with hardnesses of up to 700 HV being experienced with higher carbon contents.

This can be made use of in the process of electron beam hardening which is discussed in section 13.2. The regions of highest hardness are however restricted to the narrow fusion and heat affected zones such that they should not be judged in the same way as in other welding processes (Fig. 88). The hardened zones are surrounded to such an extent by ductile material that for example a transverse bend test can be carried out across the weld and an angle of 180° achieved. Thus in the case of welded components subject to third party assessment the general rule that all hardnesses should be below 350 HV can be replaced with bend tests or fatigue and internal pressure tests together with metallographic assessment [78]. If steels with particularly high carbon contents are to be welded such that there is a risk of cold cracking, then preheating and post-weld heat treatments must be carried out. Depending upon the geometry and thickness of the component, a de-focused electron beam with suitably adjusted beam power can be used to carry out these heat treatments.

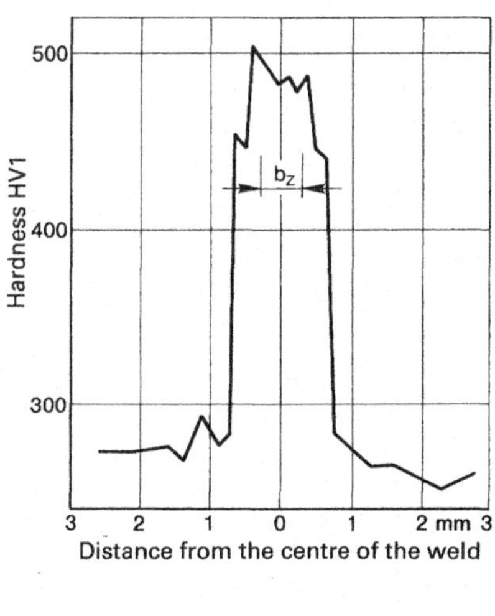

Fig. 88 Hardness trace across a weld in 21 CrMoV 5 11. t = 2.5 mm, v_S = 150 mm · s^{-1}, not post heat treated.

With unalloyed and low alloy steels, particular attention should be paid to ensuring low phosphorus and sulphur contents in order to counter the danger of hot cracking. In specific cases it may be necessary to agree on a suitable melt analysis with the manufacturer.

A typical example of this type of behaviour is the material 15 Mo 3. In Ref. 79 it has been reported that with weld thicknesses of 50 mm the tendency to hot cracking could only be suppressed once the phosphorus and sulphur contents were reduced in each case to below 0.01%.

Even high alloy steels are generally suitable for welding. In some cases, however, the tendency to hot cracking is nevertheless a particular problem. In this group of materials hot cracks form as a result of the interaction of internal shrinkage stresses and the presence of low melting point eutectics at the austenite grain boundaries. The cracks are intercrystalline and can appear as remelting cracks (subweld cracking), and in the case of electron beam welding mainly as solidification cracks. It is now generally believed, based upon many such failure investigations, that hot cracking cannot be prevented solely by the presence of a few percent of δ-ferrite. The way in which the melt has solidified now appears to be of greater importance. Depending upon the composition of the steel alloy, it will solidify either primarily ferritically and have a high degree of resistance to hot cracking, or primarily austenitically and have an increased tendency to hot cracking [80]. This boundary between the two types of solidification, which corresponds to the 'eutectic plane' at 72% Fe, 15% Cr and 13% Ni in the iron-chromium-nickel phase diagram, can be plotted on the well known Schaeffler diagram.

Here it can be seen that it does not correspond to the 0% δ-ferrite line (Fig.

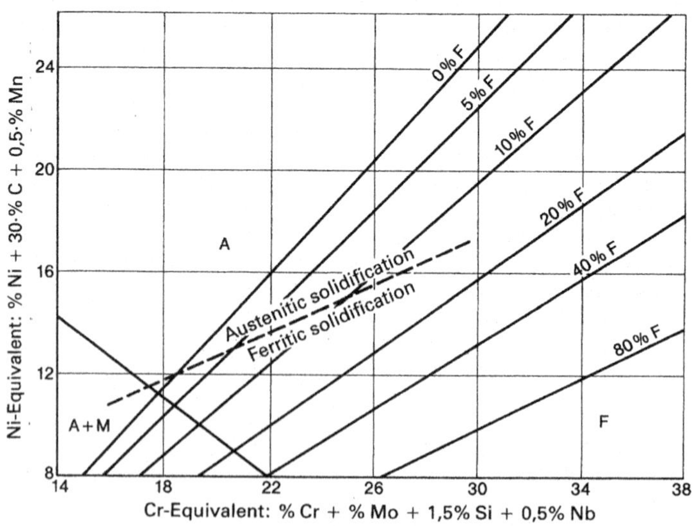

Fig. 89 Austenitic and ferritic solidification expressed as a Schaeffler diagram [80].

89). From this it follows in general that, for welding of high alloy steels, materials, and above all consumables, should be chosen which, depending upon the composition, produce a δ-ferrite component of from 0 to 15% in the melt, and which solidify primarily ferritically. A δ-ferrite component higher than 15% will on the other hand be detrimental and must be avoided. In such a case a ferrite network will be formed in the steel which will dissociate on heat treating into carbides and austenite (sigma phase) and which will be susceptible to intercrystalline corrosion.

In electron beam welding, other factors become superimposed on to these relationships. It needs to be taken into account that with few exceptions welding is carried out without use of additives and that there is no possibility of controlling the composition of the weld metal and thus the type of primary solidification and proportion of δ-ferrite formed. Apart from this the extremely high cooling rates hinder the establishment of a metallurgical equilibrium in the melt. Thus for electron beam welding, together with the localised internal stresses, it is sometimes necessary to modify the position of the limit of the tendency to hot cracking in the Schaeffler diagram [81]. However, the research material upon which this would be based is not as yet adequately confirmed statistically in order to allow reliable conclusions to be drawn.

The decrease in the tendency to hot cracking during electron beam welding has been confirmed by observations made during the welding of hardenable steels. As tests carried out on the alloy steel X 5 NiCrTi 26 15 have shown, the tendency to hot cracking can be suppressed by beam oscillation and, surprisingly, by use of lower welding speeds [82]. In hardenable steels, the grain boundary precipitation processes from the over saturated solid solution play an important role in the occurrence of hot cracking. It would appear that effects interact arising from the violent mixing of the melt caused by beam oscillation and the low internal stresses as a result of the reduced speed of welding, in that with certain weld settings the conditions for hot cracking to take place are not met.

Figure 90 shows a typical transverse hardness profile across an electron beam weld in a hardenable steel. The hardness falls to that of the level of the

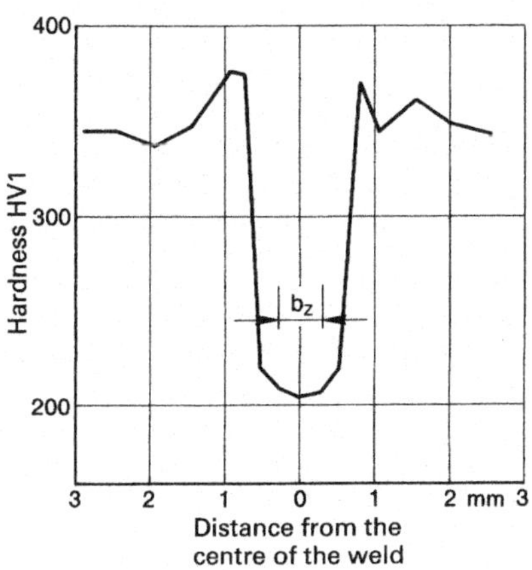

Fig. 90 Hardness trace across a weld in X 5 NiCrTi 26 15. $t = 3\,mm$, $v_S = 83\,mm \cdot s^{-1}$.

solution annealed material both in the melt and in the heat affected zones, whilst in the transition to the unaffected parent metal it increases to significantly above the initial value due to a quenching and ageing effect. Even in hardenable steels the hardness trough, as a result of its relatively small extent, does not seem to have any marked effect on the static strength properties [83]. For dynamic loading, however, it is necessary to subject the steel to a complete program of post-weld heat treatment.

As in all welded constructions, this cannot remove the low strength cast microstructure of the fusion zone which will act as a metallurgical notch [84].

Finally it must be mentioned that, in the case of high alloy CrNi-steels, in many instances the weldability is dependent both on the oxygen and nitrogen contents of the metal as well as on the stabilising elements niobium and titanium. Thus certain alloy steels, and in particular high temperature steels, will only meet the stringent requirements for successful electron beam welding if they have been vacuum melted. In the case of niobium and titanium containing steels, over stabilising should also be avoided. In other words not only should the minimum stabilising element content specification be met, but the maximum content should also not be exceeded. A number of investigations are in progress to establish suitable limits for the stabilising element contents of these steels.

Both ledeburitic steels and high speed steels are considered weldable. In contrast to quenched and tempered steels, the high carbon content does not cause an increase in hardness because, as a result of the high rate of cooling of the melt and the heat affected zones, these steels contain a large proportion of retained austenite.

Cast iron containing lamellar and spheroidal graphite, as well as malleable cast iron, are only suitable for welding to a limited extent using electron beam techniques. In these materials the weld solidifies forming a brittle and hard white cast iron microstructure highly susceptible to cracking, which can be improved by post weld heat treatment, however [85].

5.4.2 Copper materials

The high thermal conductivity of copper causes the molten metal to solidify very rapidly which thus impedes closing of the keyhole and the escape of gaseous elements. Furthermore, the greater than average thermal expansion of copper results in an increased tendency to internal shrinkage stresses. OF-coppers smelted without use of deoxidising agents and with phosphorus and oxygen contents of less than 5 ppm are however quite suitable for welding. SE-copper is deoxidised using phosphorus and thus commonly has an oxygen content of up to 30 ppm. During welding the oxygen and phosphorus compounds dissolve and, together with graphite impurities, cause considerable gas reactions to take place in the melt. These cause heavy porosity and formation of cavities which is difficult to suppress, even using beam de-focusing, beam oscillation and beam pulsing, etc. [86, 87].

Other copper based materials such as CuCrZr [88], the copper nickel alloys, and aluminium and tin bronzes are all suitable for welding. Brass, however, is unsuitable for electron beam welding because of its zinc content.

5.4.3 Nickel and cobalt materials

Pure nickel and nickel materials with low alloy contents are suitable for welding, including the well known 'Monel' alloy containing 35% Cu. Electron beam welding is now an extremely important process in fabrication of high strength nickel alloys for engine and turbine construction. With increasing alloy content, in particular of aluminium and titanium, the tendency to hot cracking increases both in the melt and in the heat affected zone [89]. The weldability of these materials can only be reliably established by welding trials using identical alloys and specimens similar in design to the actual component to be fabricated. The internal stresses produced must also be taken into account. With nickel alloys also, a decrease in the speed of welding can help to prevent hot cracking [90].

The majority of high temperature, corrosion and wear resistant cobalt alloys can also be successfully electron beam welded (Fig. 91).

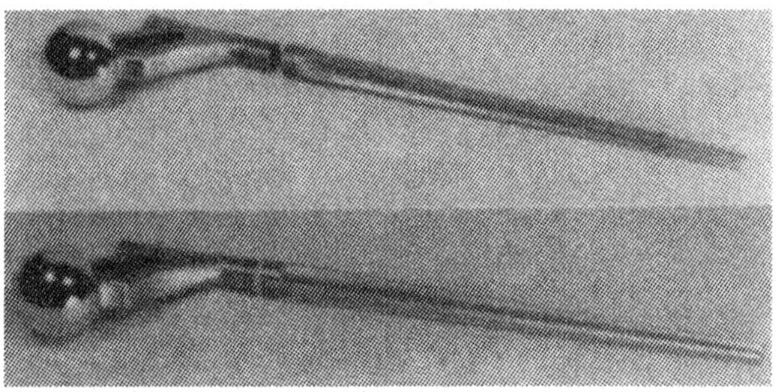

Fig. 91 Hip prostheses made from a cobalt alloy. Top – component parts, Bottom – after electron beam welding.

Stellites are a particular group of cobalt alloys whose relative levels of hardness are dependent upon their carbide contents. Their limit of weldability is at a hardness of 45 HRC.

5.4.4 Aluminium materials

In general in welding of aluminium the tenacious oxide skin, the high melting point difference between the oxide (2050 °C) and the parent metal (660 °C) and the high affinity of the weld pool for atmospheric gases, in particular for hydrogen, must all be taken into account. Other welding processes overcome these particular characteristics of the material by using alternating current arcs, shielding gases, etc. In contrast, with electron beam welding no additional steps need be taken. The electron beam easily breaks through the oxide skin, melting it in the process, whilst the vacuum keeps any harmful gases away from the weld pool. Electron beam welding is thus excellently suited to welding of aluminium, and as such only a few comments on the metallurgical processes which take place are necessary.

As a supplement to Table 2, Table 3 lists additional details of the weldability of various important aluminium alloys. In the non-hardenable wrought materials a noticeable increase in the tendency to hot cracking occurs as the magnesium content increases, which finally gives way to increased vaporisation of the magnesium and the danger of increased porosity.

Table 3 The relative weldability of various aluminium alloys [91]

Alloy group	Material example	Weldability
Non-hardenable wrought materials	Al 99.5	Good
	AlMn 1	Good
	AlMg 3	Tendency to hot cracking
	AlMg 5	Good, vaporisation of Mg, tendency to porosity
Hardenable wrought alloys	AlMgSi 1	Tendency to hot cracking
	AlCuMg 2	Good
	AlZnMgCu	Not suitable, vaporisation of Zn and Mg, porosity, hot cracks
Non-hardenable casting alloys	G-AlSi 12	Good
	G-AlSi 9 Cu 3	Good
	G-AlMg 5	Good, vaporisation of Mg, tendency to porosity
Hardenable casting alloys	G-AlSi 7 Mg	Good
	G-AlMg 5 Si	Good, vaporisation of Mg, tendency to porosity
	G-AlCu 4 Ti	Tendency to hot cracking
	GD-AlSi 8 Cu 3	Porosity

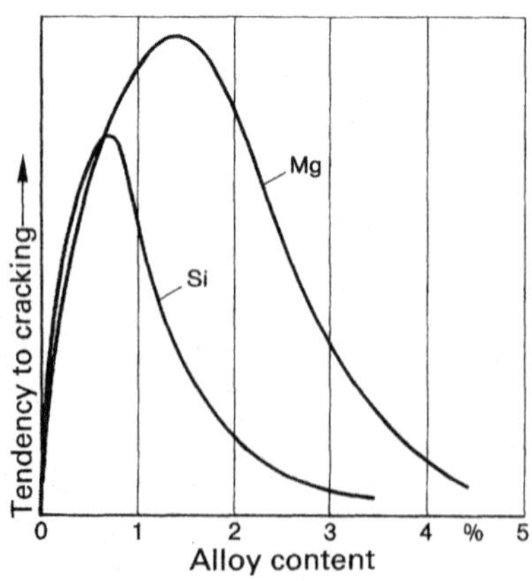

Fig. 92 Relative tendency to cracking of the alloying components in aluminium [95].

In fact the maximum tendency to cracking occurs with magnesium contents between 1 and 2% (Fig. 92 and Table 4) such that, as with methods of welding employing a shielding gas, only an increased amount of alloying (S-AlMg 5) is able to help. Apart from the simultaneous tendency to formation of porosity, the additional process and equipment requirements for welding with additives are considerable, such that in practice welding of materials such as AlMg 3 and G-AlMg 3 should be avoided. Similar tendencies to cracking are also recognised in respect of both copper and silicon in aluminium (Table 4).

Table 4 The relative sensitivity to cracking of aluminium alloying components [95]

Alloy group	Highest crack sensitivity at	Min practical content	Critical temp range, °C
AlSi	0.75% Si	2% Si	660-577
AlCu	3% Cu	5% Cu	660-547
AlMg	1.2% Mg	3.5% Mg	660-449
AlSiMg	0.5–0.8% Si 0.2–1.2% Mg	2% Si	

Amongst the hardenable wrought alloys, the excellent weldability of AlCuMg 2 must be mentioned (Fig. 93). For other welding processes with high heat inputs, this alloy is regarded as unsuitable for welding. In electron beam

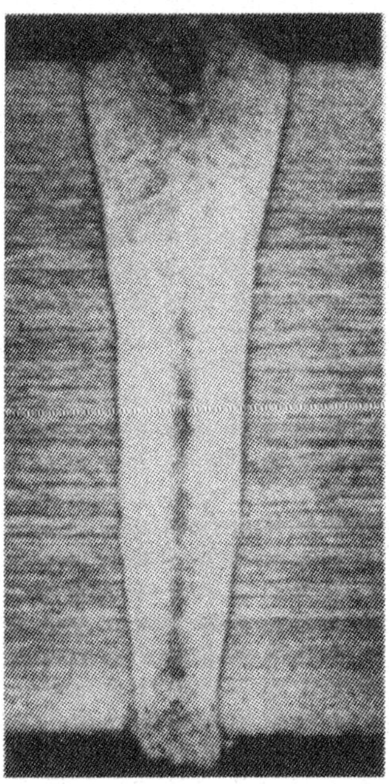

Fig. 93 Cross section through an electron beam weld in AlCuMg 2. Thickness of sheet: 7.5 mm.

welding of aluminium, zinc is regarded as an especially unwelcome alloying element. It increases the sensitivity to hot cracking, as well as causing considerable formation of porosity in the melt because of its high vapour pressure. As with other materials, and in particular with aluminium alloys, use of beam de-focussing and oscillation are not sufficient to prevent formation of porosity during welding in aluminium alloys containing zinc. International co-operative investigations have also confirmed the low weldability of the AlZn group of materials [92].

A large proportion of cast aluminium materials are also suitable for welding (Tables 2 and 3). An interesting example of this is the electron beam welding of the cooling channels in aluminium pistons (see Fig. 69) [93]. To achieve the required cast properties the aluminium materials are alloyed with silicon. Up to 13% Si (the eutectic point) the alloys have good weldability, whilst with higher Si contents the tendency to cracking increases. Furthermore, a certain degree of porosity is common in all cast aluminium materials, and to a lesser extent this is carried over into the fusion zone during electron beam welding. Particular attention must be paid to this in pressure castings alloys, in which the porosity largely depends upon the previous solidification process in the cast component [94, 95].

Work hardened and hardenable aluminium lose a certain amount of their hardness in the narrow fusion zone. Cold ageing alloys again experience an increase in hardness after welding, but do not regain the original hardness of the unaffected parent material. In the heat affected zone, however, other effects occur. Within the solution annealing temperature range the hardness increases because of a kind of 'warm' ageing, and in those areas heated to temperatures between the solution annealing and ageing temperatures over-ageing takes place and becomes apparent as a loss in hardness (Fig. 94).

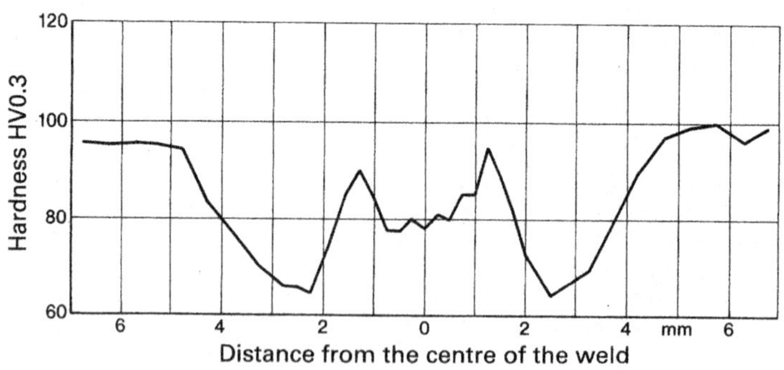

Fig. 94 Hardness trace across a weld in AlMgSi 1 [91]. t = 10 mm, b = 2 mm.

5.4.5 Titanium materials

Titanium is a highly valued material due to its low density ($\gamma = 4.5\,g \cdot cm^{-3}$) and high strength ($R_{p0.2} > 1200\,Nmm^{-2}$), and has already been electron beam welded in thicknesses up to 260 mm [96]. Titanium alloys tend to be used in applications involving high dynamic loadings and thus the very highest degree of preparation of the workpiece is required to ensure a complete absence of defects in the welds (see Fig. 115 and 116).

Titanium is one of the group of materials sensitive to gases. It reacts with the atmospheric gases oxygen, nitrogen and hydrogen at relatively low temperatures of about 250 °C, losing its ductility. As a result of this sensitivity to gases, gas shielded welding processes require use of special additional pieces of apparatus such as extended gas shields or argon filled working chambers [97]. In contrast, in electron beam welding a pressure of $p_A = 7 \times 10^{-4}$ mbar ensures an environment virtually free of residual gases which cannot be achieved with any of the normal shielding gases and which prevents occurrence of any tempering colours.

With few exceptions, all the titanium alloys are very suitable for electron beam welding. The only exceptions are the β-alloys which tend to become brittle in their heat affected zones because of metallurgical effects.

The low thermal conductivity of titanium causes marked grain growth in the fusion and heat affected zones such that during electron beam welding high speeds are preferred. Furthermore, a sharp decrease in the solubility of gases, hydrogen especially, must be taken into account during solidification since this further increases the tendency to porosity. In addition to the well known methods of reducing porosity, for example increasing the size of the weld pool and improving degassing (beam de-focussing, beam oscillation, etc), particularly careful cleaning of the faying surfaces, preferably by pickling, also helps to prevent porosity (Table 6). Titanium alloys are equally weldable in either the solution treated or aged condition. The components are annealed in a vacuum furnace to reduce internal stresses (Table 5).

Table 5. Stress relief annealing of titanium materials (from information sheet DVS 2713)

	Stress relief annealing	
Identification	Temperature, °C	Holding time
Ti 1; Ti 2; Ti 3; Ti 4 Ti 1 Pd; Ti 2 Pd; Ti 3 Pd	450–550/L	15 min–2 h
TiAl 6 V 4	500–600/L	3 min/mm, min 30 min, max 4 h
TiAl 6 V 6 Sn 2 TiAl 4 Mo 4 Sn 2	500–600/L	3 min/mm, min 60 min, max 4 h

5.4.6 Special metals

Amongst the special metals are materials such as beryllium, molybdenum, niobium, tantalum, vanadium, tungsten and zirconium, all of which like

Table 6. Recommended methods of cleaning various materials for electron beam welding

Material group	Cleaning treatment (vol %)		Pickling temperature	Ref. source
Al	Milling, planing, turning, etc. without coolant, or			
	Etching	50% H_2O 38% HNO_3 12% HF	RT	[107]
Be	Etching	75% H_3PO_4 20% H_2O 5% H_2SO_4 7 g CrO_3 per 100 g etching soln	Boiling	[99]
Cu	Etching	93% H_2O 7% HNO_3	Boiling	[107]
		subsequently, 80% H_2O 10% $K_2Cr_2O_4$ 10% H_2SO_4 or, 100% HNO_3	RT	
Mo	Etching	95% H_2SO_4 18 g/l Cr_2O_3 4.5% HNO_3 0.5% HF	RT	[99]
Nb	Etching	55% H_2O 22% HF 15% H_2SO_4 8% HNO_3	RT	[99]
Ni	Etching	47% HNO_3 32% H_2SO_4 21% H_2O	RT	[107]
Ta	Etching	90% HF 10% HNO_3	RT	[107]
Ti	Degrease	Acetone, alcohol or MEK, no chlorine containing solvents		[99]
	Etching	70% H_2O 29% HNO_3 1% HF	<50°C	[99]
X6CrNiTi 18 10	Milling, planing, turning, etc.			
Zr	Degrease	Acetone, alcohol or MEK, no chlorine containing solvents		[99]
	Etching	62% H_2O 32% HNO_3 6% HF	RT	[107]

titanium are of particular importance in certain branches of industry, but compared with normal materials such as steel, aluminium, etc, are not often welded. In many respects these special metals behave similarly on welding and can thus be treated as a generic group of materials [98]. All of them have a marked tendency to become embrittled as a result of absorption of atmospheric gases even at relatively low temperatures below about 300 °C.

The possibility of electron beam welding them in a vacuum is thus particularly attractive. In welding it must be remembered that all of the special metals are carbide formers and thus require a virtually oil free vacuum. For this reason, turbo-molecular and cryogenic pumps are preferred to oil diffusion pumps. Furthermore, adequate but not excessively long pickling (danger of hydrogen absorption) should be carried out to clean the faces to be welded (Table 6).

The weldability of beryllium is not determined by its high affinity for oxygen, but by the extremely toxic nature of its metal vapour. As yet no uniform and legally binding requirements have been published for dealing with the toxic beryllium vapour emitted into the working chamber and vacuum pump unit during electron beam welding. A close working relationship with the health and safety executive and other responsible authorities is essential to establish the necessary safety requirements. Furthermore, the hexagonal lattice structure of beryllium, with few slip planes, results in formation of coarse grains of low ductility in the fusion and heat affected zones of the weld. With increasing speed of welding, coarsening of the grains decreases but the tendency to undercutting and porosity increases. Beryllium workpieces greater than 1 mm thickness must thus be pre-heated to about 400 °C and welded with a defocused beam or using beam oscillation [99, 100].

Molybdenum and tungsten also form coarse grains on welding and thus have a tendency to embrittle (Fig. 95). To avoid cracking when welding molybdenum it is advisable to provide I butt joints with a centering lip (Fig. 96), to increase

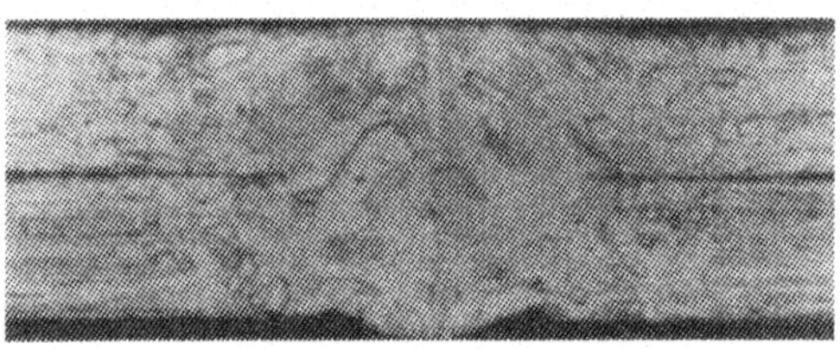

Fig. 95 Electron beam welded lap joint between molybdenum sheets. t = 0.1 mm, v_S = 150 mm · s^{-1}.

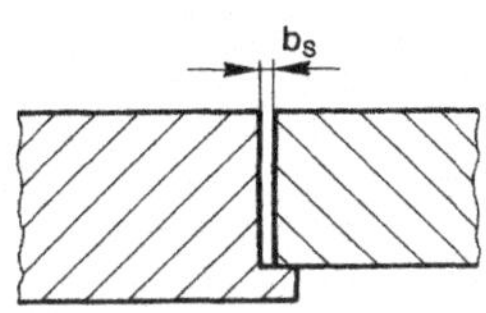

Fig. 96 I butt weld with centering lip and enlarged weld gap, as preparation for electron beam welding of molybdenum workpieces.

the width b of the weld gap, and to preheat the workpiece to at least 1000 °C before welding. This high preheating temperature requires use of furnaces specially protected with ceramic tiles, which also prevent the welded component from cooling too rapidly [101]. In addition, the weldability of molybdenum can be improved considerably by the addition of alloying agents such as rhenium (Mo 41 Re), titanium and zirconium (TZM). These elements form solid solutions having low rates of grain growth and good ductility [102]. In the case of tungsten, however, the coarse and brittle microstructure of the weld can only be refined by hammering or rolling at about 1000 °C under vacuum, which is of course seldom possible in a real manufacturing situation. In such a case special care must be taken in the structural design to ensure that the weld is only subjected to relatively low loads.

Zirconium and its best known alloys Zircaloy-2 and Zircaloy-4 both exhibit weldability as good as titanium. With standard electron beam welding technology and a working pressure of $p_A < 2 \times 10^{-4}$ mbar, porosity free welds retaining the alloys' corrosion resistant properties can be produced. Niobium, tantalum and vanadium are also highly suitable for electron beam welding.

5.4.7 Dissimilar material combinations

Even when the respective weldabilities of two different materials are known, how they will behave when welded together still cannot be predicted. Not only do their differing physical properties such as melting points and coefficients of thermal conductivity and thermal expansion play an important role, but also the metallurgical behaviour of both partner materials acting as a binary alloy in the fusion zone. In the literature numerous tables indicate the weldability of different combinations of materials in a much simplified form. Figure 97 shows a more exact representation, which sub-divides the weldability into three levels and further gives a detailed explanation of the materials evaluated.

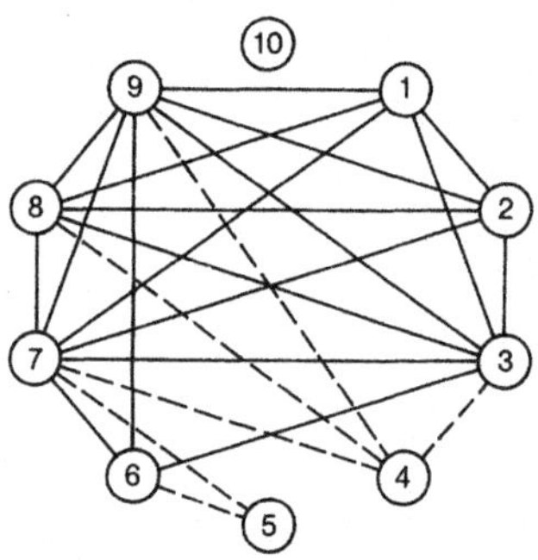

Solid lines = Suitable for welding together; Broken lines = Suitable for welding together under certain conditions; No line = Not suitable for welding together

No.	Material Group
1	Unalloyed and low alloyed low carbon steels, in particular structural steels, not free machining steels.
2	High alloy steels with high carbon contents, in particular ledeburitic tool steels.
3	High alloy very low carbon steels, in particular corrosion resistant steels.
4	Cast irons with lamellar or spheroidal graphite, black tempered cast iron.
5	Aluminium alloys with the exception of those containing zinc.
6	Pure copper, low alloy copper.
7	Copper alloys with the exception of those containing large amounts of zinc (brass, nickel silver).
8	Nickel and nickel alloys.
9	Cobalt alloys, with the exception of hard materials containing large amounts of carbides.
10	Titanium and titanium alloys.

Fig. 97 Suitability of various combinations of materials for electron beam welding [103].

The main difficulty with many molten mixtures is formation of highly brittle intermetallic phases. Whether or not a combination of materials is likely to form intermetallic phases depends upon the position of the base elements in the periodic table. To be able to make a prediction, the elements in the periodic table are rearranged with the metals divided into two separate groups

N	O	P	Q	
22 Ti	40 Zr	72 Hf	90 Th	Group B
23 V	41 Nb	73 Ta	91 Pa	
24 Cr	42 Mo	74 W	92 U	
25 Mn	43 Tc	75 Re	93 Np	Group A
26 Fe	44 Ru	76 Os	94 Pu	
27 Co	45 Rh	77 Ir	95 Am	
28 Ni	46 Pd	78 Pt	96 Cm	
29 Cu	47 Ag	79 Au	97 Bk	

Outer electron shells

Fig. 98 Elements listed according to their electron distribution to determine their relative suitability for welding in combination with one another [104].

(Fig. 98). The metals in group A will mix together forming perfect solid solutions and are thus readily weldable in combination. For the metals in group B, this is only partially the case. The group A metals form intermetallic phases with the group B metals and so they are not suitable for welding together.

This is also the case for combinations of group A or B metals with other metals not listed in either of the two groups, such as aluminium, beryllium, magnesium, etc. In this scheme, details of 46 of the most important metals together with the weldability of over 1000 potential material combinations are stored in a computer data base [104].

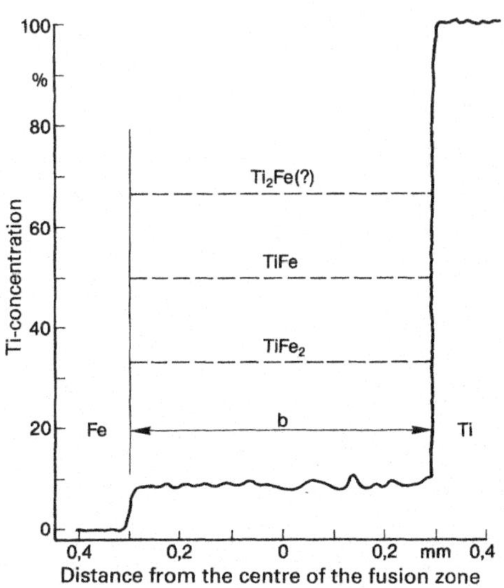

Fig. 99 Concentration of titanium in the fusion zone of a weld made with beam offset, also showing the position of various intermetallic phases found in the Fe-Ti equilibrium phase diagram [107].

In electron beam welding, however, formation of certain intermetallic phases can be avoided. As the phase diagrams of many combinations of materials show, certain intermetallic phases form preferentially at medium concentrations, for example between 33 and 66% in the iron-titanium diagram (Fig. 99). Concentrations outside this range are free of intermetallic phases, however, and thus generally ductile. If during welding the electron beam is directed preferentially to one side of the weld gap, but such that the more distant material is still melted, then this gives a method of controlling the composition of the melt as required and thus achieving welds of high ductility [105,106]. To keep variations in concentration in the melt as small as possible, such welding should be carried out using a pulsed electron beam. Other intermetallic phases, for example those forming between aluminium and titanium, are suppressed solely by the typically high rates of cooling occurring in electron beam welding.

The difficulty in transferring these laboratory findings into actual manufacturing practice is in maintaining the close tolerances required. These are necessary to ensure both the required concentrations in thicker workpieces throughout the whole depth of the weld, as well as the high accuracy of lateral movement of the workpiece with respect to the electron beam across the width of the weld.

The electron beam can also be shifted to such an extent that whilst one partner metal melts completely, the surface of the other is only wetted. In this case a soldered joint is made with the addition of the molten metal from the molten partner. This solution is particularly advantageous where the partner metals have very different melting points and the electron beam is positioned closer to the metal with the lower melting point. To ensure uniform wetting

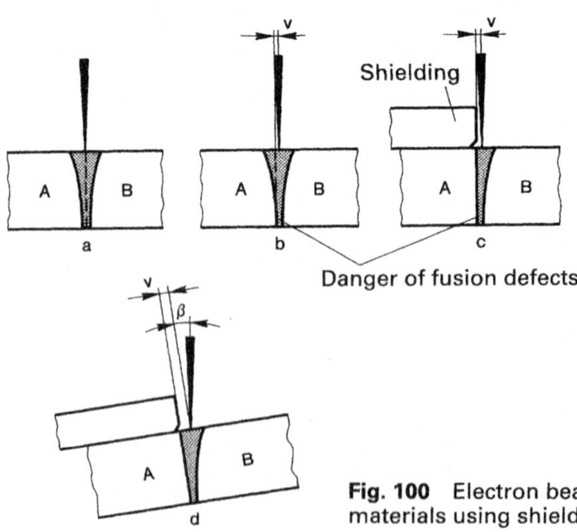

Fig. 100 Electron beam welding of a combination of materials using shielding and tilting of the workpiece [105].

rather than partial melting, even in the wider upper part of the fusion zone, use of shielding together with deflection of the electron beam or tilting of the workpiece is recommended (Fig. 100). The many successful trial welds produced using this technique show that formation of brittle intermetallic phases can thus be prevented in numerous different combinations of materials [107].

It is also possible to weld dissimilar materials successfully using additives. These become the third partner in the weld, and are added either in the form of a buffer layer which is welded to the other partners separately, or in the form of an intermediate layer or foil or a continuously fed wire which is melted together with the other two partners (see section 6.9). Although the metallurgical effect on the fusion zone is one of the most interesting aspects of electron beam welding with additives, to date very few basic investigations on this theme have been carried out [108]. One of these investigations concerns welding of ferritic to austenitic steels using various additives [109]. In this case also, welding was carried out with beam deflection, but formation of chromium carbides simultaneously with areas denuded of chromium could only be suppressed by addition of a high nickel containing additive. It also showed that electron beam deflection was required to prevent segregation occurring in the fusion zone.

5.5 Material related beam deflection

Electron beams are affected by both electrical and magnetic fields, which in particular instances can result in both unexpected and disruptive beam deflection.

Thus, for example clamping arrangements and other welding aids to be used close to the position of welding should either be made from non-magnetic materials (aluminium, austenitic steels, bronze, etc,) or in the case of ferritic steels should be demagnetised before use. This is particularly the case with workpieces which have been clamped magnetically during machining before welding. Furthermore, no electric motors or magnets (safety switches) should be positioned close to the beam path. Non-metallic materials (for example plastics) must be covered with an earthed metal foil (for example aluminium) to prevent build-up of a static charge.

In addition to these factors, the residual magnetism present in thick ferritic steel workpieces may also cause the electron beam to be deflected unexpectedly. The high residual magnetism not only stems from the chemical composition and heat treatment of the material, but also from the method used to machine it. Residual magnetism leads to a magnetic flux density distribution in the region of the weld which varies during the course of welding and deflects the beam randomly in different directions. Measurements have shown that the magnetic flux distribution above the workpiece has a considerably greater effect on the electron beam than that within the workpiece [110]. From this result come two recommendations for preventing beam deflection by the workpiece.

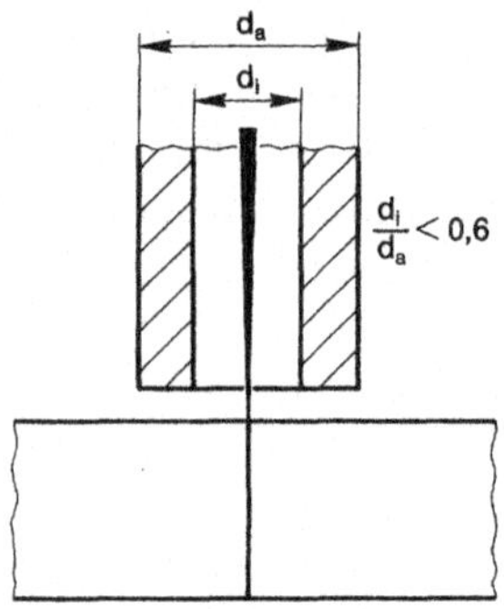

Fig. 101 Shielding of the electron beam against deflecting voltages set up by the material being welded.

- De-magnetise the workpiece to a value of below $\pm 1\mathrm{A} \cdot \mathrm{cm}^{-1} \approx 1.25$ Oersted.
- Above the surface of the workpiece, magnetically shield the electron beam in a metal tube (Fig. 101). The tube should be made of a high permeability material and, taking into account the optical observation required, should have an internal to external diameter ratio of less than 0.6 [110]. With larger wall thicknesses, use of de-magnetising should be combined with use of a shielding tube.

Finally, beam deflection can also occur during welding of dissimilar metal combinations. This is caused by thermal stresses between the different materials and by differing temperatures. Here also the electron beam wanders from its normal axis above the workpiece, but in this case can be suitably steered back into the plane of the weld [111] (Fig. 102). Deflections within the partner materials can be compensated for by application of additional external magnetic fields. The high cost of this is however seldom justified in practice [112].

In this respect, Ref. [113] advises that ultrasonic testing be carried out during welding to test for the presence of fusion defects and misplaced welds resulting from beam deflection.

For example, using this technique, deviations between the fusion boundary

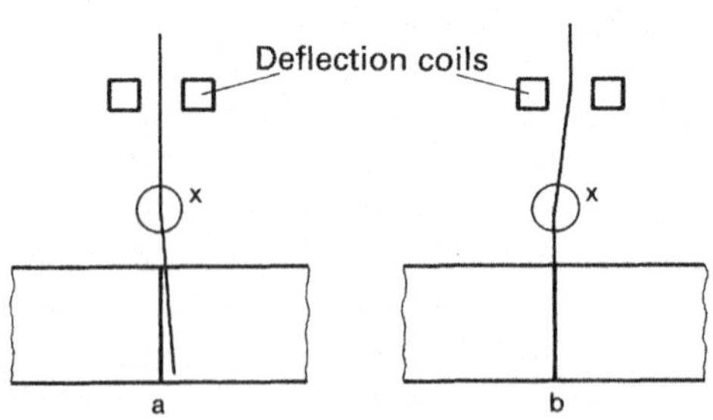

Fig. 102 Compensation of beam deflection caused by the workpiece at x, by the application of an opposite deflecting voltage.

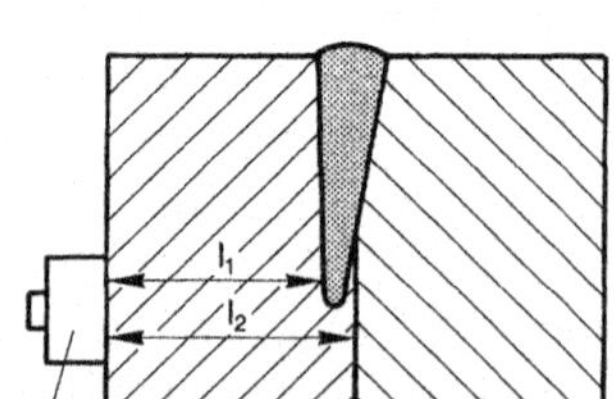

Fig. 103 Ultrasonic testing during electron beam welding to locate the possible presence of fusion defects [113].

and the butt preparation ($I_1 - I_2$ in Fig. 103) as small as 0.6 mm can be measured at a fusion depth of 30 mm. Nevertheless, this method of testing is only of limited use as the ultrasonic probe itself can deflect the electron beam, and testing becomes impossible if a root run is also welded.

6 Preparation of the workpiece

6.1 General

In welding technology it is well known that gas flames and arcs are only able to melt metal by conduction to a few millimetres of depth because of the physical limitations of their low power density. Thus as a result of use of this type of heat source only small volumes of molten metal flow together to solidify and form the weld seam. In the case of butt welds, where the thickness of the metal is greater than the depth of material which can be melted, it is thus necessary to machine the edges of the workpiece to permit the heat source gradually to melt the entire thickness of the workpiece (multipass welding technique). This also requires use of filler metal to make up for the metal previously removed by the machining operation. From the economic point of view this is a very unsatisfactory method of working. Only in welding of fillet welds does addition of filler metal add to the strength of the component to any great extent.

In contrast, preparation of workpieces for electron beam welding is carried out for completely different reasons. As a result of its considerably higher power density, the electron beam, as described in detail in section 3, is capable of melting material to a depth of several centimetres and allowing it to flow together to form a solidified weld of the same depth. The usual V or X machining of the edges of the workpiece is thus not required, nor is a filler metal (Fig. 104).

This considerably simplifies preparation of the workpiece and makes possible methods of metal joining that no other welding process can offer. Taking into account also that, in comparison with other welding processes, electron beam welds are characterised by their very low distortion, it is understandable why electron beam welding is used with such economic success for joining finished machined components. The prerequisite for this is that the particular requirements for electron beam welding must be designed into the workpiece early in the development or construction phases of the product [114].

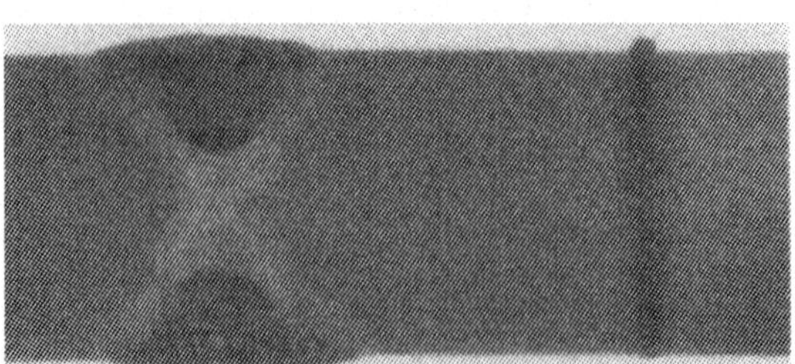

Fig. 104 Comparison of the cross sections of a multipass weld, left (manual metal arc weld), and an electron beam weld, right in material of identical thickness. t = 25 mm.

A number of joint designs and their most important characteristics are discussed below to assist the design engineer in choosing a workpiece preparation most suitable for the particular requirements of the component. Figures 105–107 show a number of the terms most often used in workpiece preparation as applied to the various types of electron beam weld. In all the sketches of welds the reader should note that for clarity the various items shown are not always reproduced to scale.

6.2 Surface of the workpiece, surface cleaning

The normal rules of fusion welding applied to cleaning of the surfaces to be joined also apply in electron beam welding. Thus, attention should be paid to the compatibility of the cutting fluids used during the machining operations prior to welding (turning, milling, etc) with the numerous metals suited to this type of welding. Only chilled iron grit should be used for blast cleaning the surface of the metal, and on no account carborundum. Surface grinding is unsuitable for soft materials such as aluminium, copper, etc, and should thus be restricted to high strength steels. Pickling using a process specially adapted to the material in question is also advantageous before many welding operations (Table 6 – Page 84).

This should be carried out in the following manner:

- Solvent degreasing, most effectively in an ultrasonic unit;
- Pickling in different acids, depending upon the type of metal: to prevent absorption of hydrogen this process should be of limited duration;
- Rinsing with distilled water;
- Drying in clean, preferably warm air;
- Storage (only for short periods) until welding.

Apart from the electron beam welding, a number of other factors must be taken into account, as in many cases it is finished machined workpieces which are to be joined.

Thus any carburised, nitrided, anodised or phosphated layers, or cadmium or zinc coatings must be removed in the region of the joint prior to welding. Diffusion products, inclusions, oxides and surface coatings produce pores and other defects in the fusion zone which can have a severe detrimental effect on the quality of the weld (Fig. 108), and should thus be removed both from the faces to be joined and from along the seam (Fig. 109) using the methods of machining mentioned previously. Further details are given in section 6.6.

6.3 Basic joint shapes

6.3.1 Butt welds

Figure 110 shows the preparation for the simplest type of butt weld. Because of the relatively small amount of machining required this type of weld is the ideal

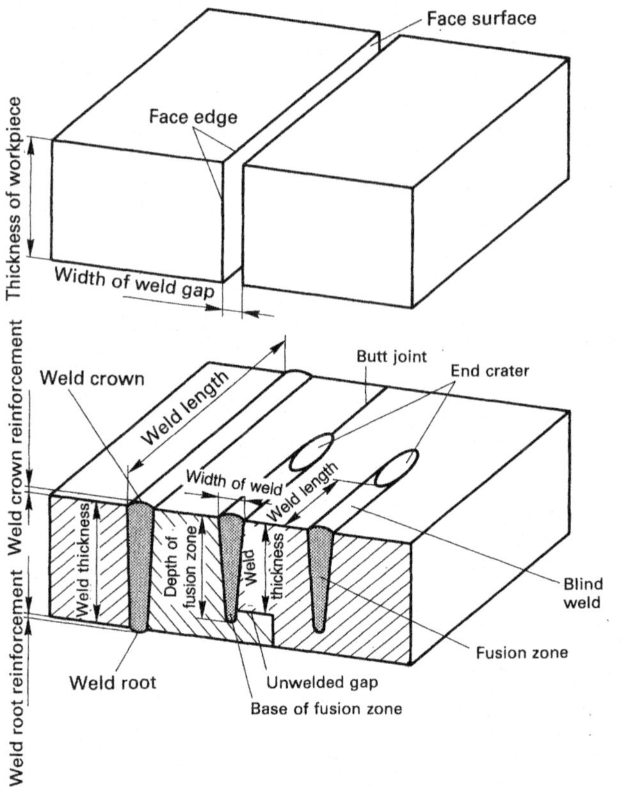

Fig. 105 Terminology used for the various types of weld preparation and weld seam.

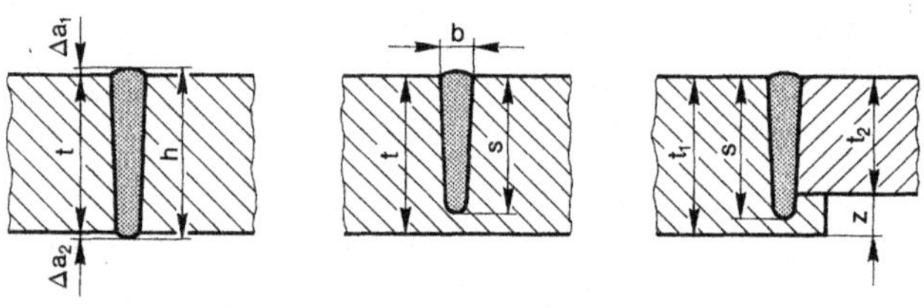

Fig. 106 Symbols used to designate weld dimensions.

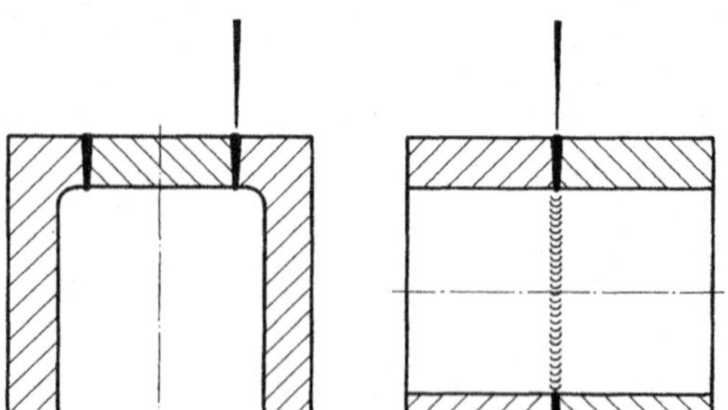

Axial weld: The electron beam is parallel to the axis of rotation of the workpiece

Radial weld: The electron beam is parallel to the radius of the workpiece.

Fig. 107 Definition of axial and radial circumferential welds.

Fig. 108 The surfaces of the crown of electron beam circumferential welds in case hardened toothed gears (4 : 1, Reproduced: 0.7 X): *a)* Case hardened layer not removed before welding; *b)* Case hardened layer removed before welding.

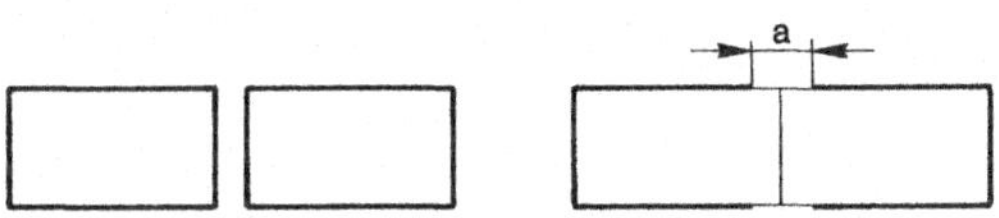

Fig. 109 Preparation of surface treated workpieces.

example for almost all types of electron beam welded joint. When free of defects, the butt weld, because of its simple geometry, is the preferred type of weld for all applications involving dynamic loading or corrosive attack. In this version, however, it lacks a keying element, such as a centering lip, to improve positioning of the components relative to one another (Fig. 111). Furthermore, for possible high strength requirements it permits only limited machining of the surface or root of the weld to be carried out, since damage to the adjacent surfaces of the workpiece with respect to the normal finishing tolerances must be avoided.

With a centering lip, as shown in Fig. 111, spatter which otherwise occurs on welding through the underside of the weld is prevented. Electron beam weld spatter is far less adherent to the surface of the metal than that produced by other methods of welding. Its removal, nevertheless, requires adequate access to the underside of the weld. If this access is prevented (e.g. in hollow bodies) but freedom from spatter has been specified because of the function of the component, then an edge preparation with a centering lip, for example, should be considered.

When designing a centering lip the dimension z should be especially carefully chosen. On the one hand the thickness of the material t must be completely welded by attaining an adequate depth of fusion zone, on the other hand all variations in depth of the fusion zone must be contained within the thickness of the centering lip z. For manufacturing and economic reasons the dimension z cannot be arbitrarily chosen, and as a guide should be $z = 0.2t$, min 2 mm,

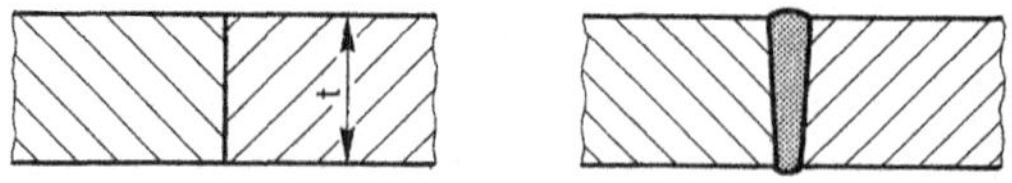

Fig. 110 I butt weld, normal version.

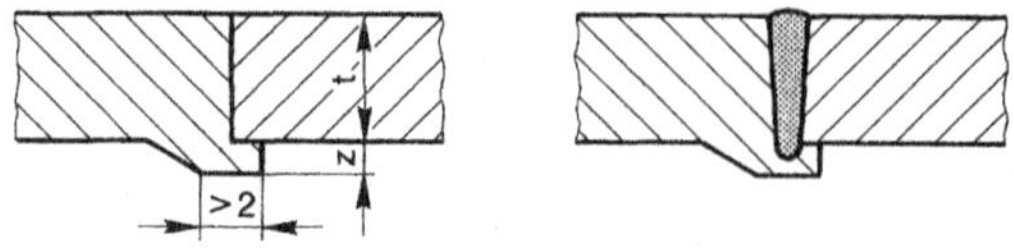

Fig. 111 I butt weld, with centering lip.

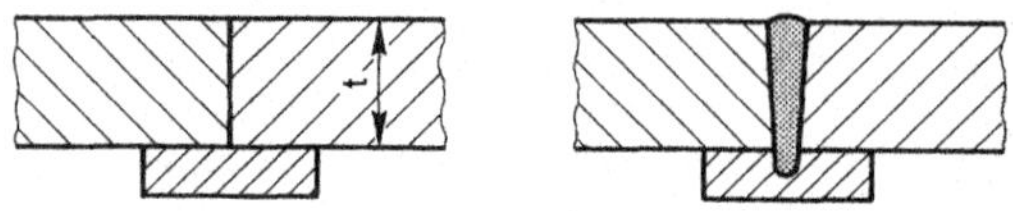

Fig. 112 I butt weld, with backing strip.

max 5 mm. The optimum thickness should be determined by trial welds on specimens of the same thickness and material as the actual workpiece.

As with backing strips often used in other welding processes, the centering lip shown in Fig. 111 and backing strip shown in Fig. 112 also both help to support the molten material in the weld pool.

As explained in section 4.3, the electron beam gives up most of its energy in penetrating the thickness of the workpiece. However, it must emerge from the underside of the weld still possessing a significant inherent energy, to ensure complete welding of the full thickness of the material in spite of variations in the depth of the fusion zone, and to exclude the possibility of root cavities and spiking. When welding simple butt welds the keyhole must also be open on the underside.

In welding butt joints with centering lips or separate backing strips these irregularities at the base of the fusion zone are accepted but restricted to a region outside the load carrying cross section of the component, and if necessary can be removed later by machining. If this is not carried out, then compared with simple butt joints, these welds will have the disadvantage of having less resistance to corrosion and impaired dynamic strength properties. In this case, the weak point is the unwelded gap between the load carrying cross section of the component and the centering lip or backing strip. At this point preferential crevice corrosion can occur, or cracks can be initiated and grow into the cast microstructure of the fusion zone.

To prevent these points of weakness, instead of removing them by machining other weld preparations are possible, for example if there is a need for centering of the workpiece. For centering rotating components during welding, a butt weld incorporating an internal or hidden centering lip can be chosen (Fig. 113). With this solution attention should be paid to positioning of the lip with respect to the upper and thus wider part of the weld seam, together with accurate positioning of the electron beam. In components where a double pass weld is intended, the weld gap between the faces to be joined should be as large as

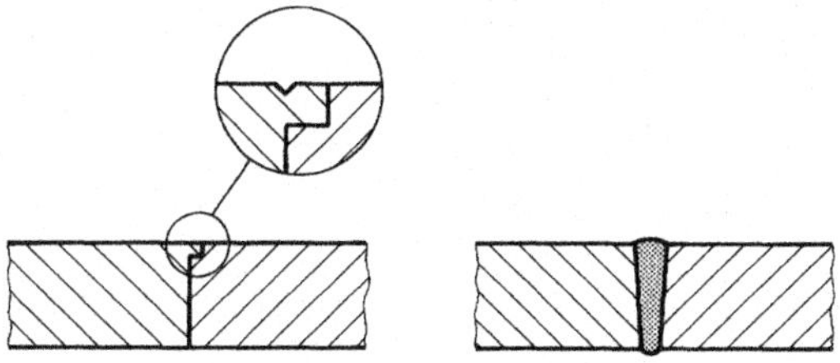

Fig. 113 I butt weld with internal centering and guide marks for positioning the electron beam.

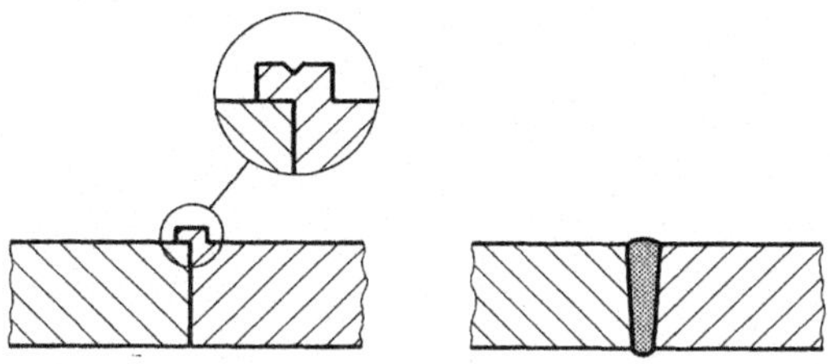

Fig. 114 I butt weld with external centering and guide marks for positioning the electron beam.

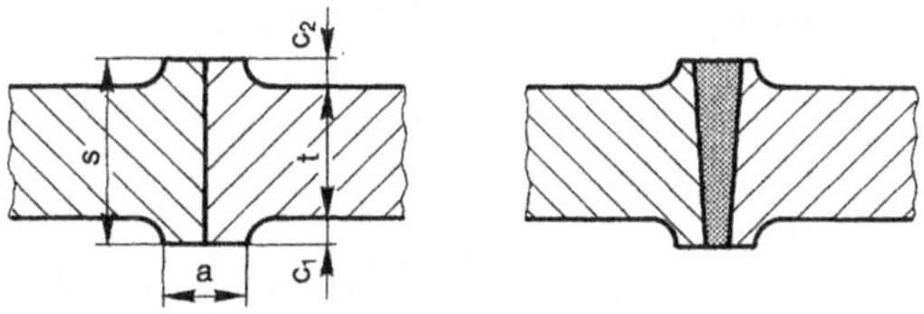

Fig. 115 I butt weld with additional allowance for machining.

possible beneath the centering lip, whilst above the lip a slightly wider gap may also be chosen.

Another possibility for butt weld preparation with centering is shown in Fig. 114. The centering lip here only partially fulfills the function of that shown in Fig. 113. Its dimensions and thus volume are kept so small that it can be fused completely and incorporated into the weld as filler material.

If centering of the workpiece components is of secondary importance, for example because this function can be fulfilled by use of a jigging arrangement, and where welds are being made which are to be highly stressed, the weld preparation shown in Fig. 115 has proven successful in practice.

In this design the crown and root of the weld are removed by machining after welding to produce a weld free of notches and undercut, as well as to provide the conditions required for the various non-destructive testing procedures. If defects are found in the weld which require welding over again (remedial welding), then using the dimensions shown in Table 7 it would again be possible to machine away the crown and root of the weld.

Figure 116 shows a method of preparation which is essentially a combination of a butt joint with additional material for machining, and a centering lip. This was developed mainly for circumferential seams in drive shafts and is characterised by a particularly small centering lip (Table 8).

Table 7 Dimensions (mm) relevant to Fig. 115

	Steel			Titanium alloys		
s	a	c_1	c_2	a	c_1	c_2
<10	12	1.0	1.8	12	1.0	1.5
10 . . . 20	12	1.5	3.0	12	1.5	2.5
20 . . . 40	12	2.0	4.8	12	2.0	4.0
40 . . . 50	12	2.5	6.0	12	2.5	5.0

Table 8 Dimensions (mm) relevant to Fig. 116

t_2	a	c
1 . . . 3	2	2
4	2.5	2
5	2.5	2
6	2.5	3
7 . . . 15	3.5	3

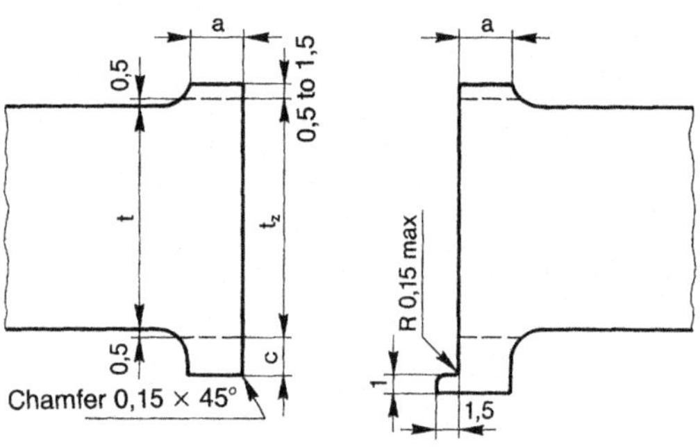

Fig. 116 I butt weld with additional allowance for machining and centering lip [115].

6.3.2 T-butt and other types of joint

If two plates are to be joined at right angles to one another by electron beam welding, then a T-butt weld as shown in Fig. 117 can be used whereby welding can also be carried out from both sides to overlap at the center of the plate. In making these welds the electron beam must be deflected at least by an amount equal to its aperture angle or the workpiece tilted to obtain free access to the weld position. It should also be taken into account that, if the angle of deflection or tilt is too large, then the faces to be welded may not be completely melted (danger of fusion defects).

It is further indicated in Fig. 117 that in electron beam T-butt welds the dimension a almost completely disappears. The reason for this is quite clear, since with few exceptions such electron beam welding is carried out without use of filler metal. From this it follows that this type of welded joint with its notch-

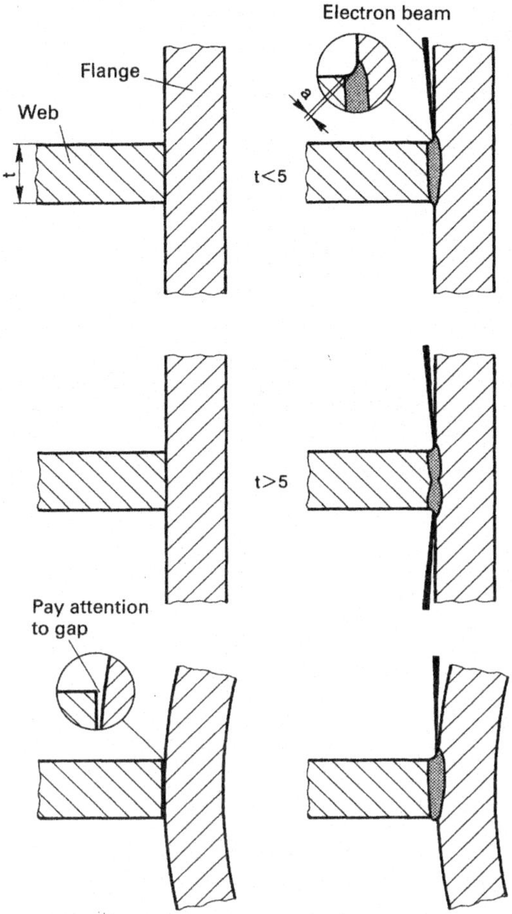

Fig. 117 T-butt weld.

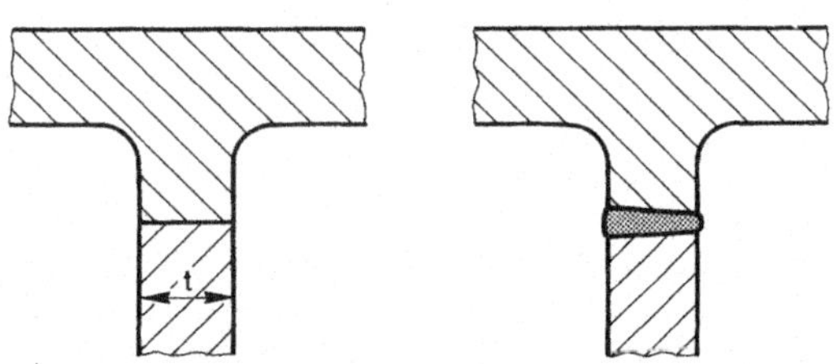

Fig. 118 T joint with I butt weld for higher strength requirements.

like shape cannot be used in applications where high dynamic stresses may be encountered. In such a case the T-joint should be modified as shown in Fig. 118 to permit welding of a normal in-line butt weld.

For less stringent strength requirements, however, electron beam welding offers a number of other interesting possibilities for making T-joints. Beginning with the principle of a basic lap weld (Fig. 119), the T-joint can welded through the flange plate directly into the web plate (Fig. 120), or with thin material into

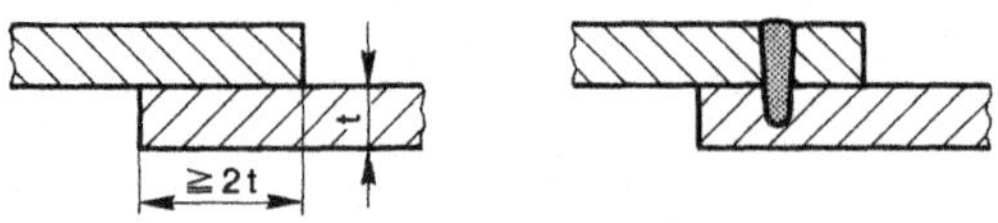

Fig. 119 Lap weld.

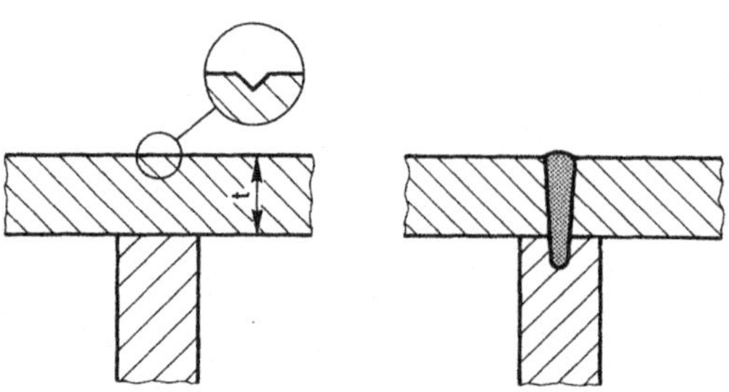

Fig. 120 T joint with I weld for low strength requirements.

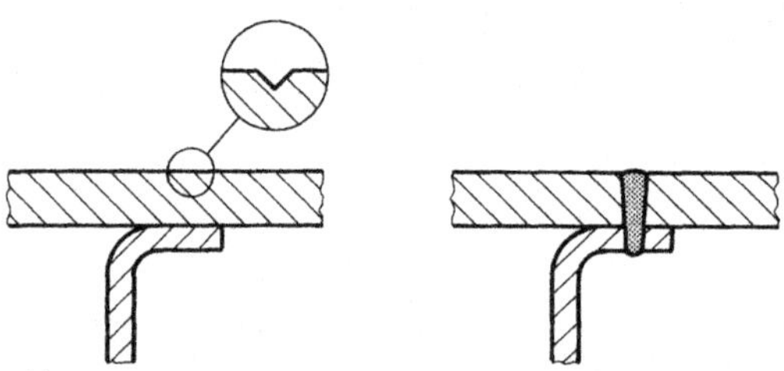

Fig. 121 Lap T joint with I weld.

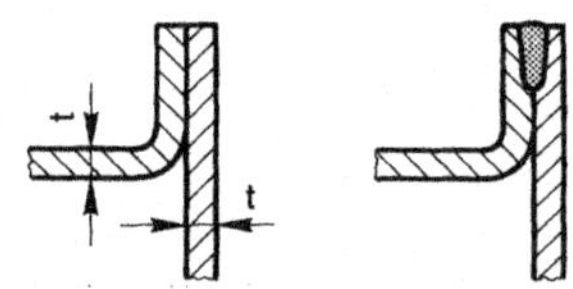

Fig. 122 Edge weld.

the end of the web plate which is bent accordingly (Fig. 121). According to the geometry of the component, edge welds can also be welded (Fig. 122). In all these examples, in spite of the very small amount of machining necessary, the narrow and deep fusion zones produced by electron beam welding are excellently suited for welding T-joints. With this type of weld, however, for strength applications, careful attention should be paid to unwelded parts of the gap.

6.3.3 Particular points regarding circumferential welds

The types of weld described so far can all in principle be used also for welding circumferential seams, so that here mention need only be made of certain particular aspects. In welding radial seams a method of centering the components is an absolute necessity. As mentioned in section 6.3.1, there is a whole series of types of preparation which can be used to achieve this depending upon the shape of the workpiece and its method of assembly and function (Fig. 123). Only in rare instances can centering of the components be achieved by use of jigs and clamps (Fig. 124). The design engineer should also not forget that careful placing of the radial weld can also beneficially affect the manufacturing costs involved (Fig. 125).

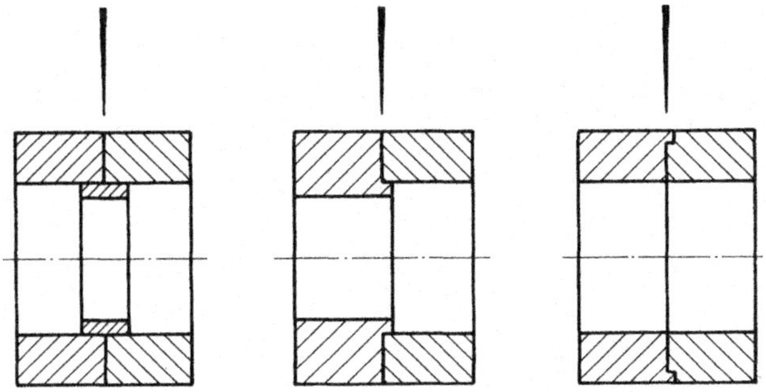

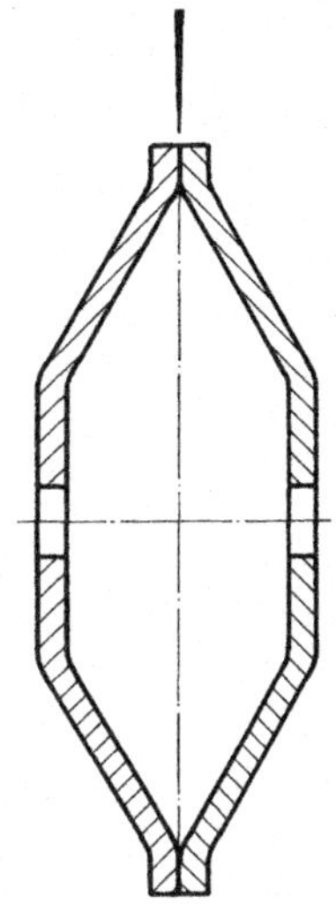

Fig. 124 Workpiece with radial weld, centered using a welding jig.

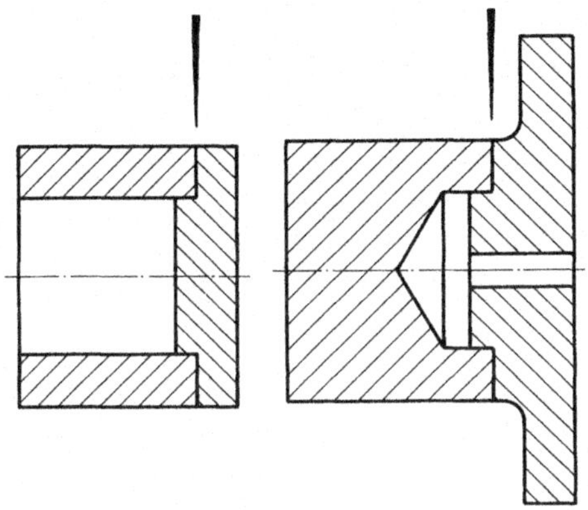

Fig. 123 Various types of weld preparation for radial circumferential welds with centering.

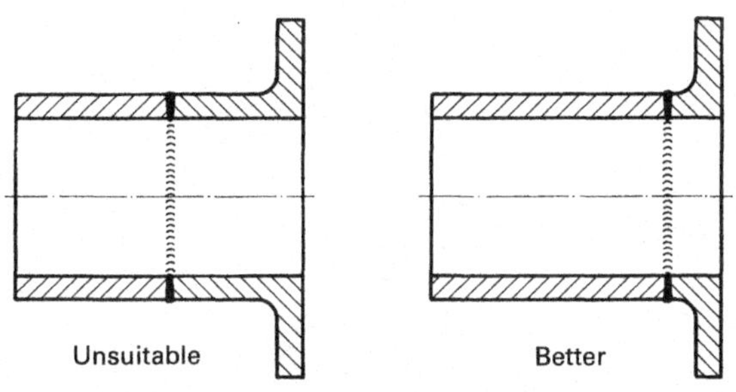

Fig. 125 Unsuitable and suitable positions for a radial weld seam.

In most cases when welding axial seams, centering can be achieved without a need for additional aids because of the location provided by the interconnecting shape of the turned components (Fig. 126). Certain tolerances must be met, however, to counter the increase in gap width caused by transverse shrinkage of the weld. More on this aspect appears in section 6.5.2.

Turned components, especially toothed gears, are designed taking into account not only their strength requirements, but also their stiffness.

Here, the situation often occurs that in the region of the weld the material cross section present need not be fully joined by welding (Fig. 127). As shallow a depth of welding as possible should suffice for transferring the torque and

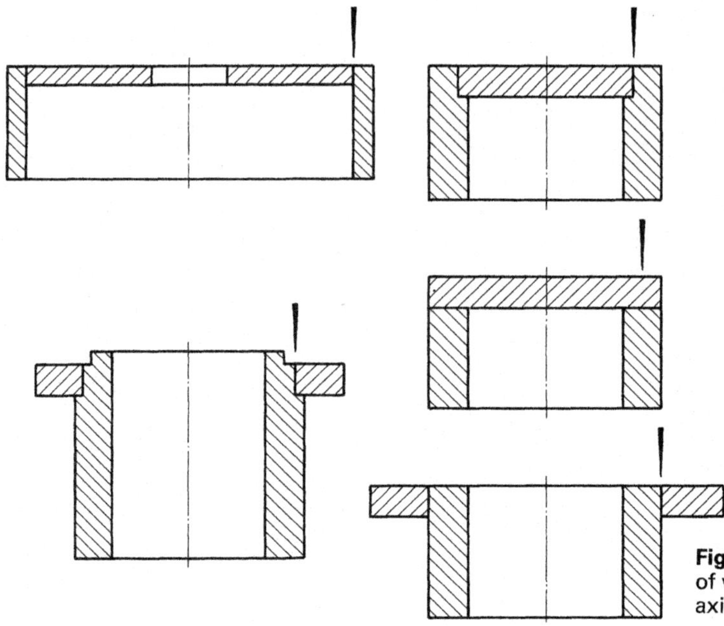

Fig. 126 Various types of weld preparation for axial welds.

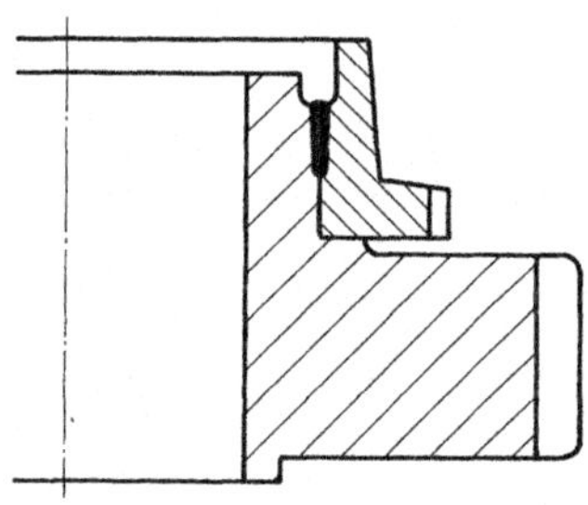

Fig. 127 Example of an axial weld with a depth of fusion zone adapted to suit the strength requirement (not welded through the full thickness of the joint).

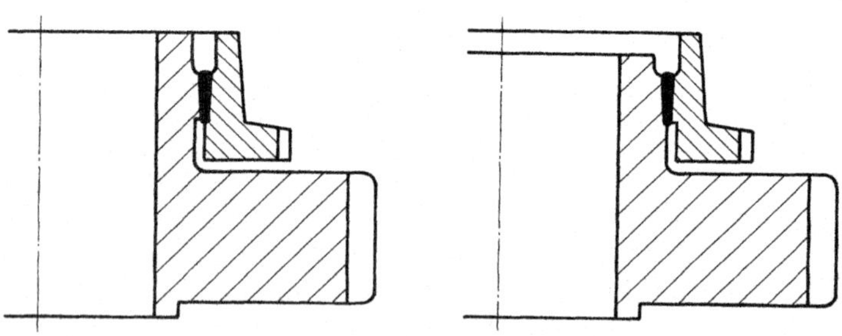

Fig. 128 Improved weld preparation for welds which are not to be welded through the full thickness of the material.

also result in a significant cost saving. Only partially welding the faces to be joined, however, results in an unwelded gap being left which, as already mentioned a number of times, is a definite disadvantage in the case of dynamic loading (Fig. 111). In addition there is the added danger that on welding the axial seam, transverse shrinkage may cause a crack to initiate in the fusion zone at the base of the unwelded gap. This can be prevented by providing a cavity at the base of the fusion zone (Fig. 128), thus forming a normal butt weld. This cavity can be formed by machining one or both of the faces to be welded, and can be left open or closed. Additional steps should be taken to evacuate closed cavities (see section 6.10). For welding large series of components, turning back the component on one side only is preferred.

6.4 Differing thicknesses

Welding together component parts of differing thicknesses without specific design considerations being met cannot be recommended for a number of reasons. With high strength requirements, the abrupt change in thickness acts detrimentally in the same way as a notch (Fig. 129, bottom).

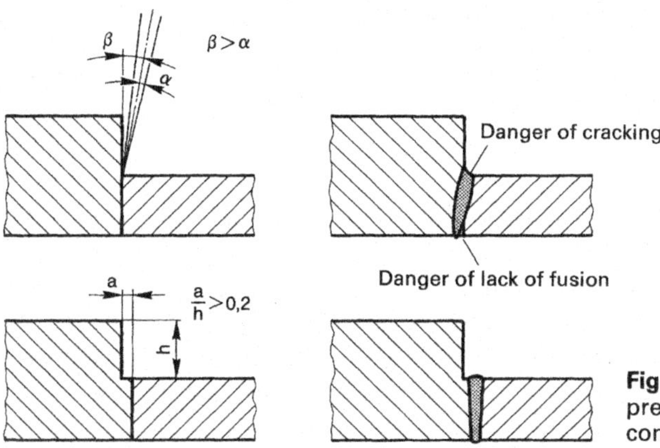

Fig. 129 Unsuitable weld preparation for welds between components of differing thickness.

There is also an increased danger of cracking in the concave region of the crown of the weld (Fig. 129, top) because of internal welding stresses, particularly in circumferential welds. Furthermore, the angle of deflection β must be greater than that of the beam aperture α_w which can lead to fusion defects in the lower part of the weld. By shifting the butt weld out of the position of change in thickness, thus equalising the thicknesses of material to be welded, the joint becomes just another normal butt weld (Fig. 130).

Fig. 130 Suitable weld preparation for welds between components of differing thickness.

In the case of larger differences in thickness, effects of unequal thermal conduction can be countered by shifting the position of the beam towards the thicker of the two components or by provision of a slot next to the butt joint (Fig. 131). This method of joining has been shown to be particularly useful

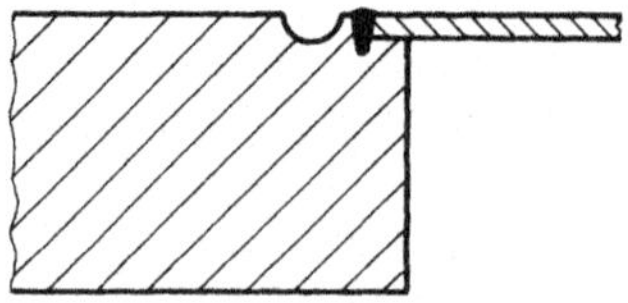

Fig. 131 Weld preparation including a slot (thermal break) between the components to be joined.

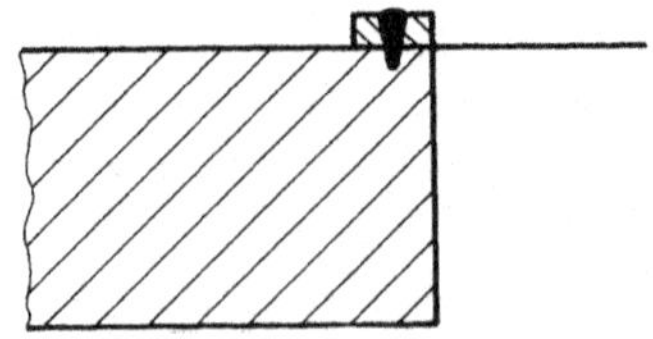

Fig. 132 Weld preparation for joining components with an extreme difference in thickness.

with certain crack sensitive materials and where minimum distortion is specified (see section 6.12.3).

Extreme differences in thickness, for example when joining foils to components, require use of either special clamping arrangements or a supplementary strip of thicker material (Fig. 132).

6.5 Dimensional tolerances

6.5.1 Gap width

As a rule electron beam welding is carried out without use of a filler material. Thus, because of the narrow fusion zone the faces to be welded must be carefully mated, which in the ideal situation would mean that they should touch without leaving a gap. In machining, however, specification of dimensional tolerances is essential to permit the faces to be joined to be machined only as accurately as required and to avoid unnecessary work. To this end, Fig. 133 gives guide values for the maximum acceptable gap width in butt welds in steel for use when electron beam welding in the horizontal position. These values are the result of numerous surveys of welding specialists from various branches of manufacturing, and in spite of their somewhat abstract mathematical method of representation have been well proven in practice. Depending upon the actual case in question it may of course be necessary to use narrower or wider gaps than those specified in Fig. 133. In the latter case, an additional widening of the fusion zone must be compensated for by use of a lower welding speed or by oscillating or de-focusing the beam, etc, as required, to prevent the weld pool falling through the gap and to prevent fusion defects.

These guide values are valid for all types of butt weld (section 6.3), as well as for T-butt welds (section 6.3.2) and for radial welds (section 6.3.3).

6.5.2 Clearance

The gaps required to fit components together (under-sizing) are a particular disadvantage in the case of axial welds. The transverse shrinkage which takes place at the beginning of welding leads to a direct and immediate increase in the width of the gap diagonally opposite (Fig. 134) which results in the weld pool dropping through the gap or in fusion defects, etc. Clamping arrange-

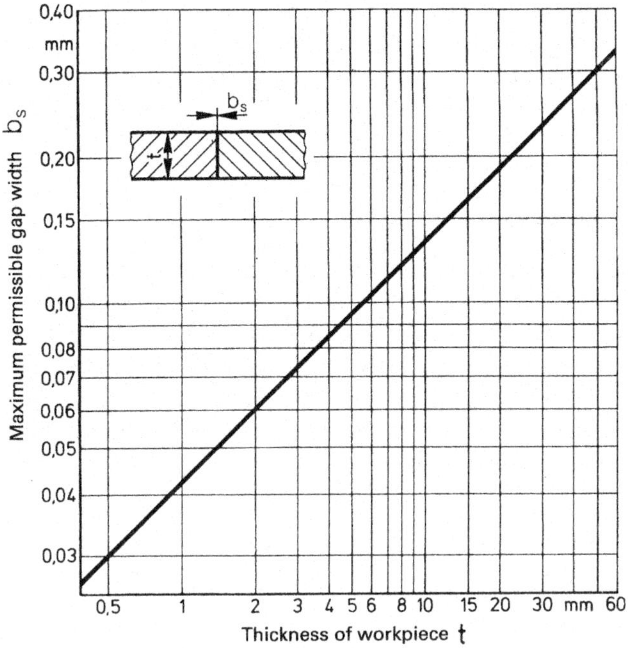

Fig. 133 Guide values for the maximum permissible weld gap for longitudinal welds made in the horizontal position in steel.

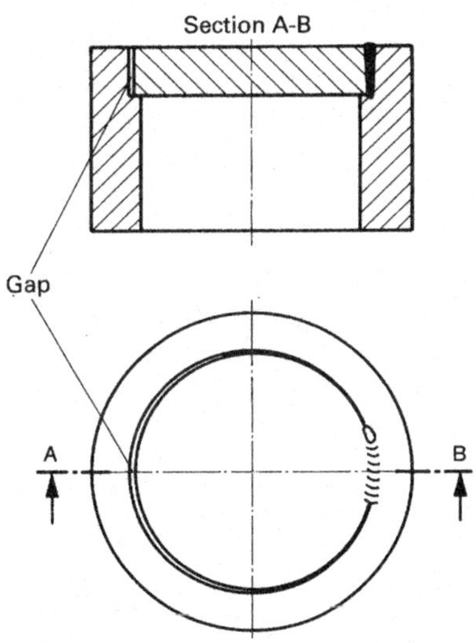

Fig. 134 Enlargement of the gap width when welding axial welds with an allowance made for clearance.

ments, as well as the use of tack welds, are only able to withstand the high shrinkage forces generated in particular cases, such as where the volume of the weld is small. Tack welding, however, requires use of an extra production step. Thus, in the widely applied electron beam welding of finished close tolerance components, such as in manufacture of toothed gears, a clearance fit should be avoided on all axial welds seams.

For this reason, in actual manufacturing practice, press fits are generally used (interference fit) with the dimension in question being very slightly oversize. Especially useful are pairs of components which can fitted together using the simplest of arrangements for welding. This is dealt with in more detail in section 6.12.3.

6.6 Machining of the weld faces

The term maximum acceptable gap width is closely inter-related with the surface roughness of the faces to be joined by welding. The machining symbol ▽▽▽ of earlier standards and the N7 or N8 of current standard DIN ISO 1302 characterise the same surface quality which corresponds to an actual surface roughness of $R_a = 1.6$ or $3.2\,\mu m$. These values are valid for all shapes of weld with the exception of lap and T-butt welds. The choice of technique for machining the surfaces is discussed in section 6.2.

6.7 Weld seam control lines

Whilst a narrow fusion zone is presented as a particular advantage of the electron beam welding process, it should not be overlooked that with inexact positioning of the electron beam with respect to the abutting faces to be welded, the danger of fusion defects increases considerably. An accidental sideways shift in the position of the beam can be caused by a number of factors such as distortion of the workpiece during welding, the presence of magnetic fields in the workpiece or thermal stresses when welding dissimilar metals.

Whilst in many cases radiographic and ultrasonic test methods can be used to locate this type of defect, because of the costs involved simpler methods of control should be employed first.

The first indication of beam deflection can be obtained by visual inspection of the weld. This requires that an increase, albeit small, can be detected in the width of the weld root because of heat build up on the underside of the joint (Fig. 135). The ability of this method to locate such defects is somewhat limited.

Another method of testing which has proven itself in practice is use of

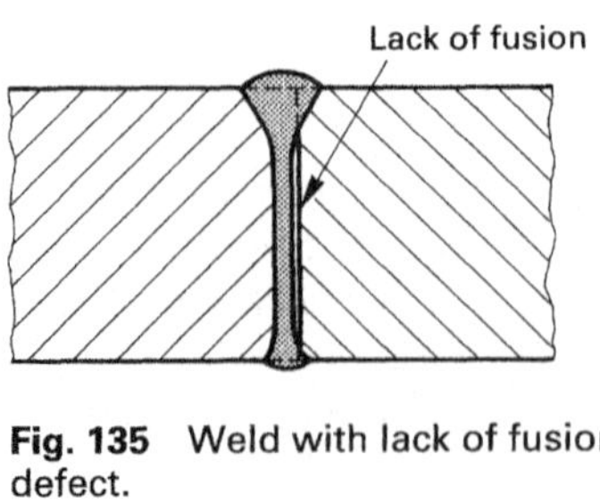

Fig. 135 Weld with lack of fusion defect.

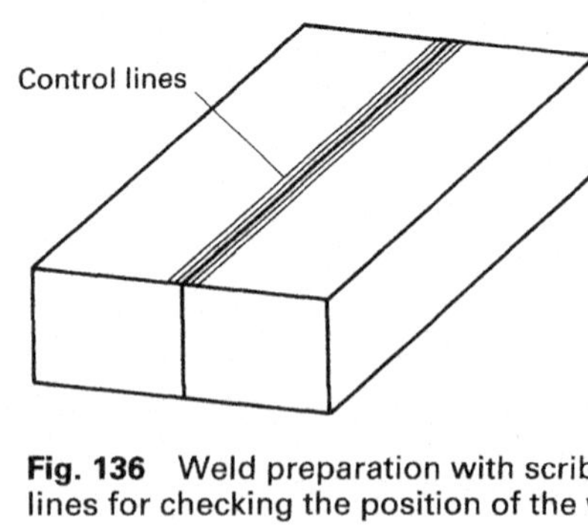

Fig. 136 Weld preparation with scribed lines for checking the position of the weld.

control lines marked on the upper, and possibly lower, sides of the workpiece (Fig. 136). These are scribed or engraved parallel to the respective edge to be welded as part of the preparation for welding, and after welding they permit the position of the fusion zone to be determined with respect to the original faces of the joint. From measurements of the width of the fusion zone within the workpiece (by macrosection), the maximum permissible beam deflection can be calculated taking into account other manufacturing tolerances.

6.8 Weld start and finish

6.8.1 General

As mentioned in section 3.3, around the keyhole transport of material takes place against the direction of welding (see Fig. 43). At the start of the weld there is thus an accumulation of fused metal in the form of a lump, whilst at the finish of the weld there is a deficit of material in the form of a crater (Fig. 137).

Fig. 137 The surface of electron beam welds showing end craters and crater cracking (1 : 5, Reproduced: 0.6 X).

Both features are considered to be weld defects or unacceptable variations in the component and as such must be avoided.

6.8.2 Longitudinal welds

When welding longitudinal welds which will not be machined at the start and finish of the seam, use of run-in and run-out plates (Fig. 138) which are subsequently cut off is recommended. These have the advantage that they also

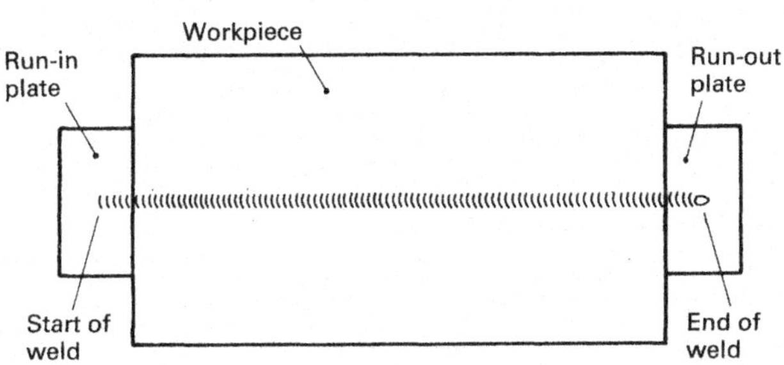

Fig. 138 Workpiece with run-in and run-out plates to enable the beginning and end of the weld to be cut off.

help to prevent a further cause of irregularity in the weld. This is the heat build-up at the edges of the workpiece which causes a kind of undercutting of the front and back edges. To prevent this, the run-in and run-out plates must be pressed, tacked or welded firmly to the workpiece to ensure good thermal contact and an efficient flow of heat away from the workpiece. The design engineer should thus ensure that additional aids are used so that the start and finish of the longitudinal weld take place away from the actual workpiece.

6.8.3 Circumferential welds

When making circumferential welds, these irregularities are prevented with the help of a particular type of beam current control, further information on which is given in section 7.2. A slow increase or decrease in the beam current ensures that the excess metal at the start of the weld and the end crater are distributed more evenly over a greater length of weld.

This is so successful that in practice the start and finish of a circumferential weld are often indistinguishable to the naked eye.

6.9 Welding with fillers

6.9.1 General

It has already been mentioned that because of the great depth of fusion achieved, the electron beam welding process does not require use of V or X preparations, so that use of a filler metal to replace the metal otherwise removed is not required. Nevertheless, a number of interesting methods of welding using fillers have been developed, amongst others in order to make fillet welds having a nominal throat dimension 'a', to fill in-line butt welds with large weld gaps and to carry out overlay and repair welding. In future, however, the main area of use of electron beam welding with filler will be in joining dissimilar or metallurgically difficult-to-weld metals (for example metals sensitive to hot cracking).

6.9.2 Design considerations

It has already been shown in Fig. 114 that by using a particular type of butt weld preparation a part of the workpiece can also be used as a source of filler metal during welding. A further example, shown in Fig. 139, is a fillet weld used for production of a circumferential joint having a defined throat dimension 'a'.

Both methods have the advantage that the filler metal added is identical to that of the parent metal and that no extra equipment is required to feed the filler into the weld pool, for example a wire feed unit.

Another possibility, shown in Fig. 140, is welding with a filler foil. Within certain limits of machining and preparation, in particular in the case of manufacturing defects (remedial welding), an over-wide weld gap can be bridged

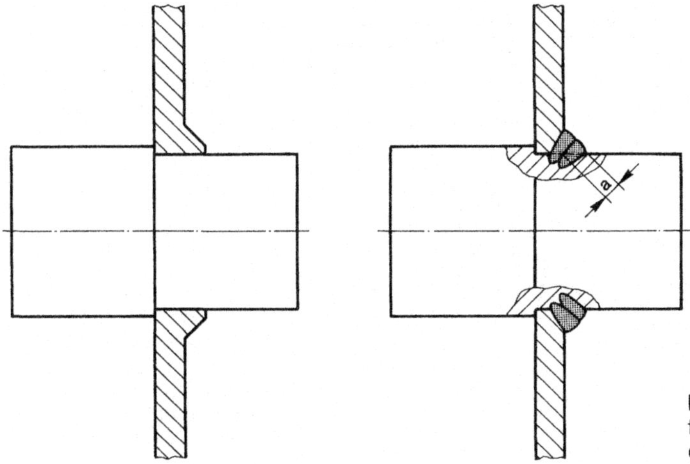

Fig. 139 Circumferential fillet weld of defined dimension 'a' in a T joint.

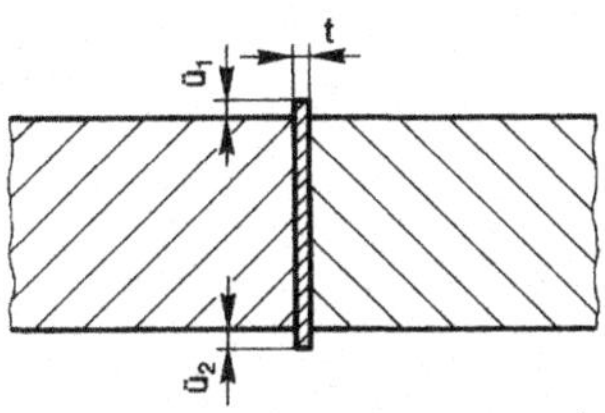

Fig. 140 I butt weld with a foil used as filler material.

and filled by incorporating an intermediate filler foil of chosen thickness t and over dimensioned by amounts u_1 and u_2. However, far greater use is made of foils of specially chosen materials for metallurgically modifying the fusion zone. With this type of butt welding the surfaces of the faces to be welded and of the filler foil must be especially carefully cleaned, since whilst the volume of the weld is increased only marginally, the surface areas involved are doubled. Use of intermediate foils is not suited to circumferential welding of machine fitted components.

6.9.3 Welding of thick walled components with fillers

Electron beam welding with a filler wire has been pursued for many years with the aim of joining poorly machined faces with varying or large weld gaps.

Successful though the initial laboratory trials were [116–118], it was shown that irregularly varying weld gaps required use of automatic wire speed control. With the advent of beam powers of over 7.5 kW, not only could thicker walled material be welded but also larger components with normal dimensional tolerances, using a mobile vacuum system. Automatic wire speed control played a particularly important role in development of this type of welding [119]. The problem of automatic wire speed control was solved by extremely rapidly sensing the unwelded weld gap using the deflected, and thus somewhat advanced, full power electron beam. At the same time, variations in the X-rays

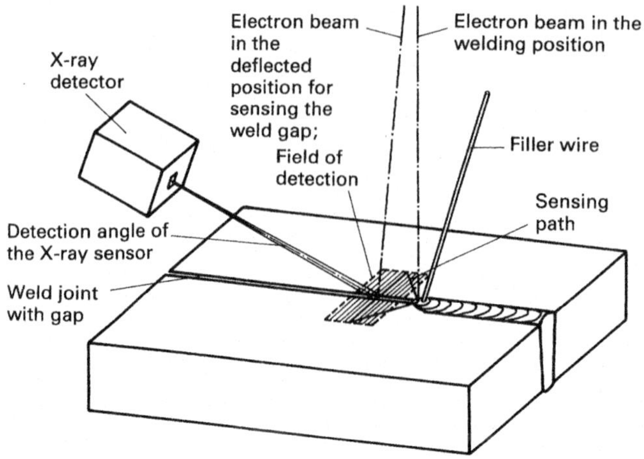

Fig. 141 Radiographic method of measurement for controlling the speed of filler wire feed in welding joints with varying weld gap width.

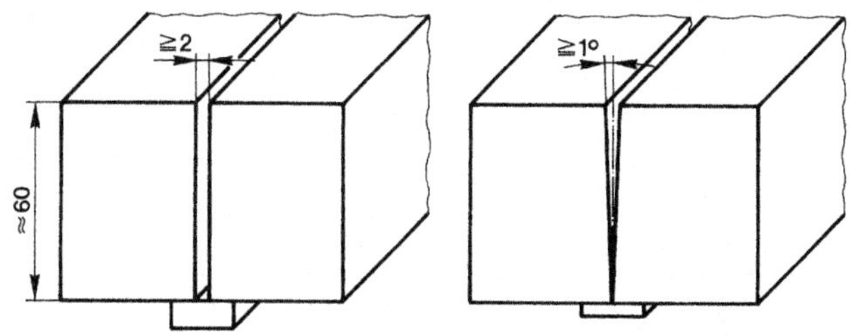

Fig. 142 Various ways of preparing the workpiece when electron beam welding thick plate materials using a multipass technique.

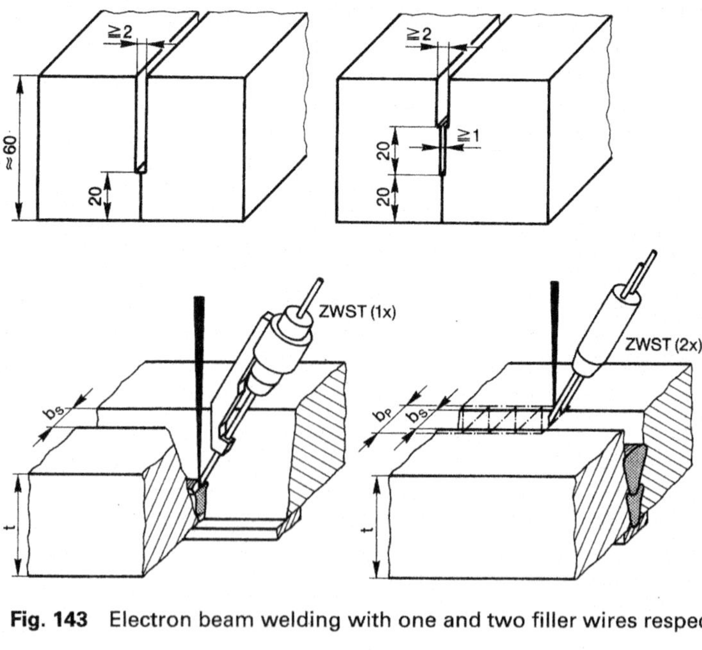

Fig. 143 Electron beam welding with one and two filler wires respectively.

generated as the beam traverses across the weld gap are measured (Fig. 141). The position of the weld gap in relation to the electron beam and the width of the gap can thus be sensed by an X-ray detector and turned into a measurement signal which can be used to control the speed of the filler wire. The instrumentation required for this type of welding, which also incorporates a number of other control and monitoring processes, is considerable and only economically justifiable when components are to be welded which, because of their size and weight, cannot be guaranteed to maintain a specified weld gap tolerance during assembly and welding. In all other cases the more normal machining of the faces of the joint is considerably more cost-effective, and for numerous electron beam welded finish machined workpieces entails no additional work above that required to ensure dimensional accuracy.

This type of welding can be used for a different purpose, however, as for example in the case of arc welding, if the faces of the joint are machined to a V or Y shaped preparation and the void left must be made up using a filler metal (Fig. 142). This type of process offers the possibility of joining components with large wall thicknesses when the beam power available is otherwise insufficient to penetrate the full thickness of the material in the conventional way in a single pass [120, 121]. Figure 143 shows schematically the arrangement of electron beam and wire feed used in the case of a V weld. It must not, however, be forgotten that a large number of additional parameters such as the speed, positioning and angle, etc, of the wire must also be taken into account. There are particular advantages to be had in using a Y shaped preparation, including a stepped increase in the width of the gap (Fig. 142, below, right). As a result of the increase in width of the gap towards the top, significantly less angular shrinkage takes place and the effective preheat applied using a multipass technique (Fig. 144) results in an overall decrease in hardness (Fig.

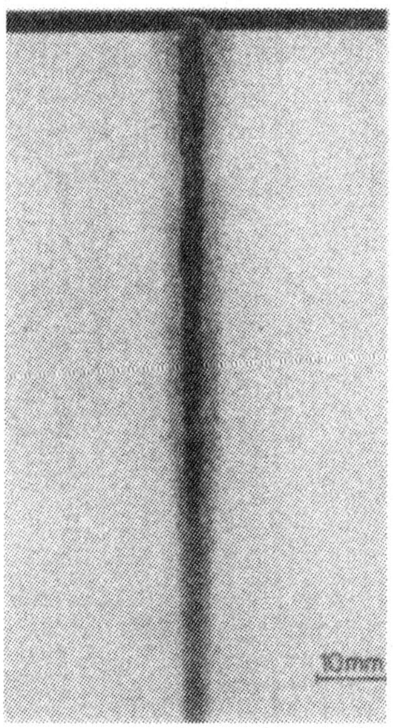

Fig. 144 Cross section through a 6-pass electron beam weld, plate thickness: 100 mm, Y weld preparation with settings $U_B = 140$ kV, $I_S = 40$ mA. 1. Layer: Without filler material; 2. to 6. Layer with 0.8 mm thick weld filler wire [120].

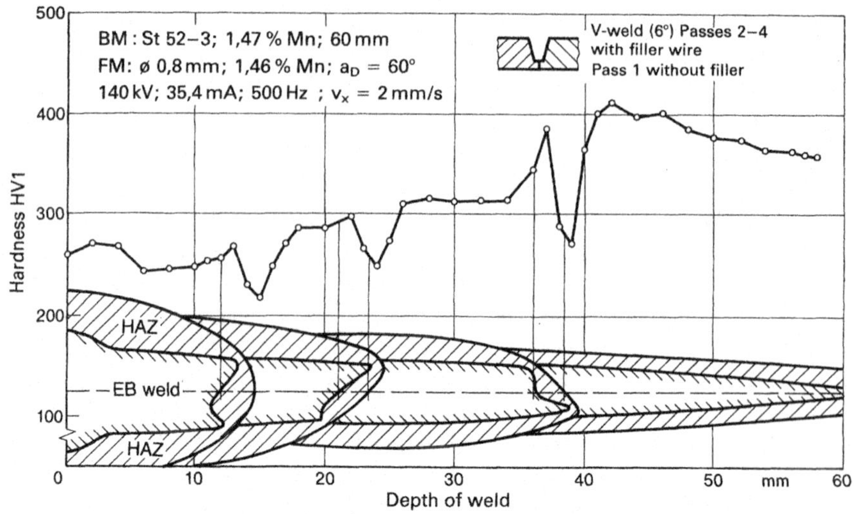

Fig. 145 Hardness profile through the depth of an electron beam weld produced using filler material, plate thickness: 60 mm, Y weld preparation with step. Parent metal St 52–3. Filler material identical, 0.8 mm thick [120].

145). Welding can also be carried out using two wires to increase the rate of melting [122]. Thus, summarising, in all cases the increased cost of electron beam welding with filler is considerable and only appears justified when there is limited beam power, large wall thicknesses are to be welded and no other solution is possible.

6.9.4 Overlay welding, repair welding

Finally, the possibility of carrying out overlay and repair welding using electron beam techniques must be mentioned. Figure 146 shows overlapping cladding runs of X 2 CrNiMo 19 12 welded on to the surface of St 37 steel. Use of parabolic beam oscillation ensured relatively slight mixing (and thus dilution)

Fig. 146 Electron beam overlay weld of X 2 CrNiMo 19 12 on St 37, U_B = 90 kV, I_S = 15 mA, v_s = 14 mm · s^{-1} [117].

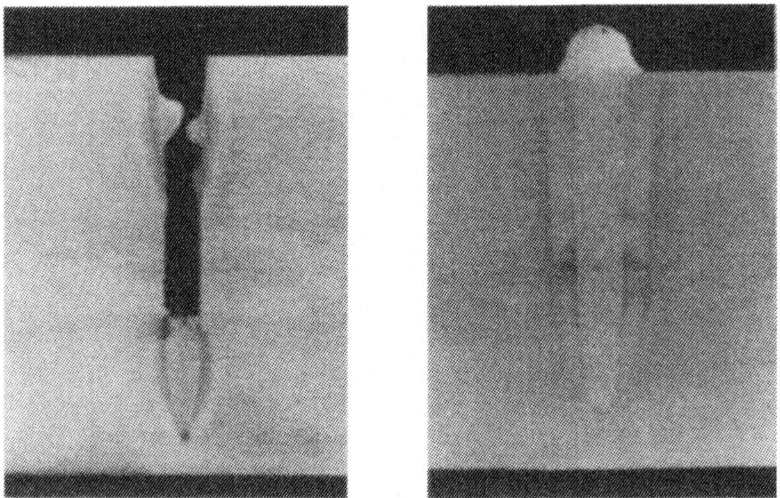

a) b)

Fig. 147 Cross section through a repair weld. Thickness of workpiece 30 mm: *a*) Slot 20 mm deep; 1.8 mm wide; *b*) As welded using filler metal [122].

of the stainless material by the ferritic base material. In repair and remedial welding, welds from which defects have been excavated can also be returned to perfect condition by welding with the addition of a filler material (Fig. 147).

6.10 Ventilation openings

If on assembling components to be welded voids are left in the vicinity of the welds, then in the absence of ventilation holes air will be trapped between the components which will not be evacuated during welding. On welding, this entrapped air will cause molten weld metal to be ejected, as a rule resulting in formation of weld defects. The cause is the difference in pressure of the entrapped air and the working chamber, which equalises once the keyhole forms and penetrates through to the void. Voids resulting from the machining operation should thus be kept as small as possible by careful design (Fig. 148). In other cases, ventilation openings such as drilled holes of diameter $d > 2.5$ mm should be provided outside the region of the weld (Fig. 149).

If ventilation holes must be drilled in the joint itself, then an area of the joint close to the end of the weld should be preferred and its cross section dimensioned such that it will be fully closed by the final part of the weld run (Fig. 150).

6.11 Welding with difficult access

Various difficult-to-access seams which cannot be reached by an arc or gas flame can be readily welded using an electron beam, as shown in Fig. 151. In

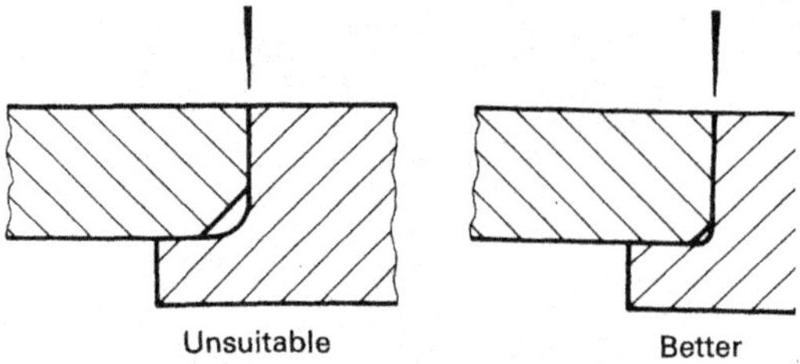

Fig. 148 Voids caused by unsuitable machining.

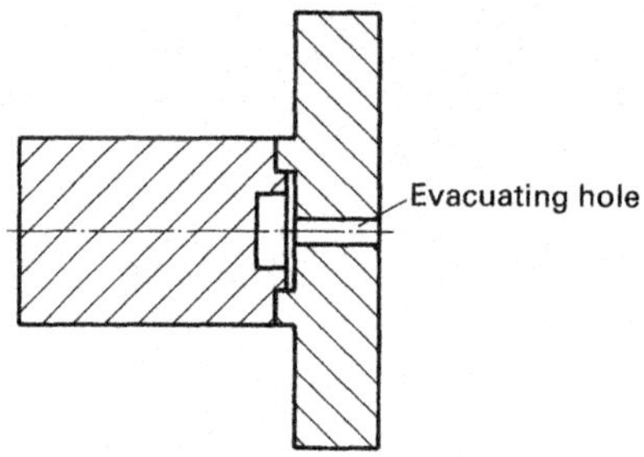

Fig. 149 Welded component with additional drilled hole to permit the space within the component to be evacuated.

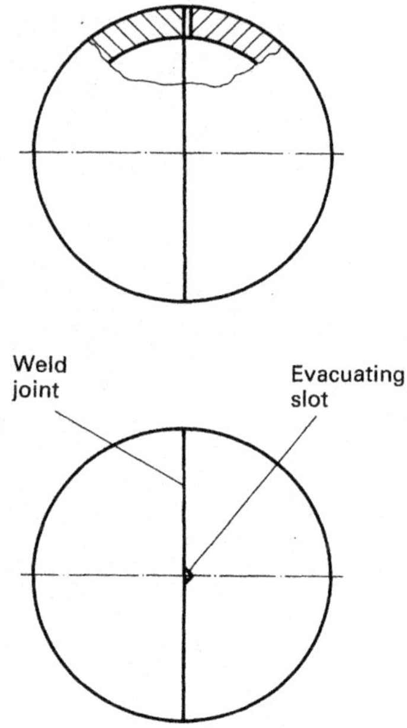

Fig. 150 Hollow welded body with evacuating slot.

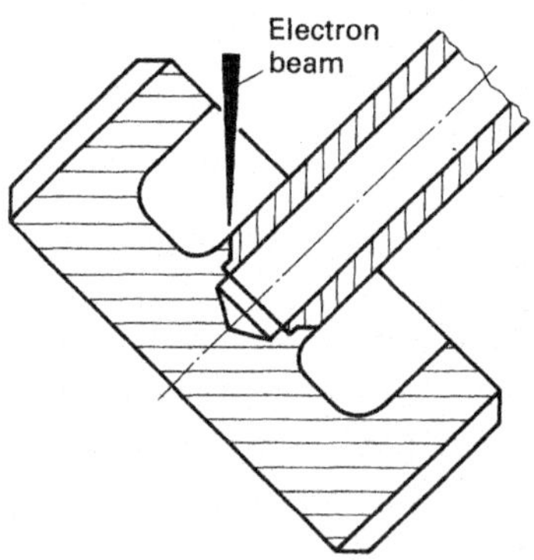

Fig. 151 Example of a difficult to access electron beam weld.

some cases even problems of completely inaccessible seams can be overcome (Fig. 152). The excess energy of the beam is used to penetrate through to the restricted or obscured welding position, and the residual energy is then used to carry out the actual weld. To position the beam accurately the surface of the workpiece must be marked accordingly. It should be remembered that these types of weld seam are almost impossible to test non-destructively, and that the blind weld seam left in the component causing the obstruction will also have an additional effect on the dimensional accuracy of the welded component.

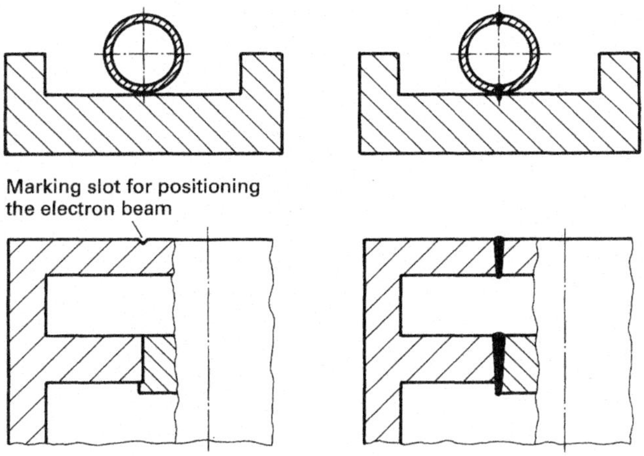

Fig. 152 Example of a seam not freely accessible to the electron beam.

6.12 Weld distortion

6.12.1 General

All welding processes cause a greater or lesser distortion of the component as a result of the laws of physics applicable to the processes involved.

In this respect electron beam welded components have the particular advantage of exhibiting extremely small amounts of distortion. In spite of this it is nevertheless necessary to be able to predict the amount and direction of this distortion in order, in practice, to be able to weld finish machined components successfully to the high degrees of tolerance specified.

The cause of weld distortion and its relationship to internal welding stresses has already been discussed in section 5.2.5.

6.12.2 Longitudinal welds

Quantitative prediction of the amount of distortion a component might experience on welding is extremely difficult because in most cases the extent to which the geometry of the component will prevent deformation cannot be estimated. Every weld shortens and narrows the component, and in the case of V shaped fusion zones also causes angular distortion. The actual amount of distortion, however, depends not on the volume of the weld but on the material, the length, width and thickness of the workpiece, and on the datum point from which it is measured. The following diagrams are a guide to the amount of transverse shrinkage and angular distortion occurring in longitudinal welds, which can be applied to actual components only where the factors involved are broadly similar.

Figure 153 shows measurements of the transverse shrinkage of 3–24 mm thickness plates of X 15 CrNiTi 18 9 steel. In each case the plates were 12 mm wide and 76 mm long. The progressive increase in transverse shrinkage which occurs as the thickness of the weld increases is clearly apparent. Figure 154

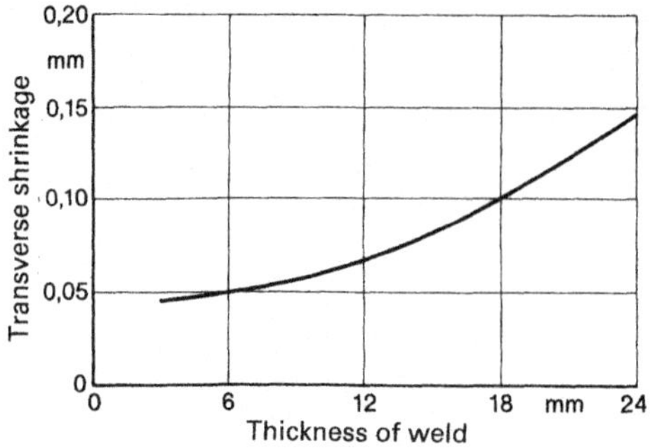

Fig. 153 Transverse shrinkage as a function of the thickness of the weld [123].

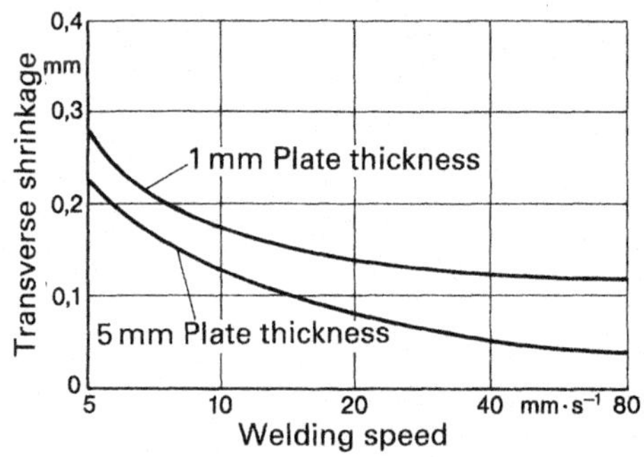

Fig. 154 Transverse shrinkage as a function of the speed of welding for two different plate thicknesses.

shows the results of another series of trials and plots transverse shrinkage as a function of welding speed. It can be seen that depending upon the welding conditions, there is a minimum distortion which cannot be further reduced. Accurate investigations have further shown that the transverse shrinkage which occurs is not constant along the entire length of the weld, but reaches a maximum at approximately its centre [124].

In the case of longitudinal shrinkage, an average value of 0.5 mm per metre of weld length may be assumed for steel. The actual value is apparently largely

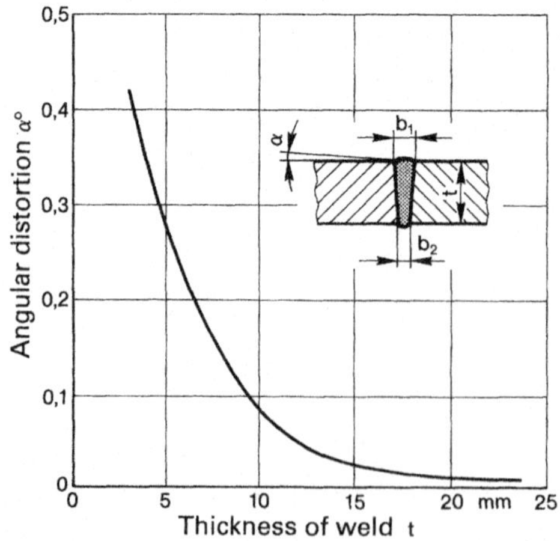

Fig. 155 Angular distortion as a function of weld thickness [123].

dependent upon the thickness and width of the weld in question. In contrast, angular distortion is strongly dependent upon the cross sectional shape of the fusion zone. The values plotted in Fig. 155 showing angular distortion were measured on welds in 3–24 mm thickness plates of X 15 CrNiTi 18 9 steel. With increasing thickness of weld the aspect ratio of the fusion zone b/t becomes smaller, that is, the fusion zone becomes less V shaped and more I shaped, which leads to a decrease in the amount of angular distortion.

Compared with single sided welds made using other welding processes, the angular distortion in electron beam welded seams is extraordinarily small even in thin material.

6.12.3 Circumferential welds

Electron beam welding has many advantages for manufacture of toothed gears, and in particular gear wheels. As the example in Fig. 156 shows, if manufactured solely by machining, the distance between the two gear crowns must be relatively large to give space for the gear milling cutter to run out. In contrast, if the gear crowns are machined separately and subsequently electron beam welded accurately together, a considerable saving in weight and material can be achieved. This type of construction is already used in manufacture of many lorry and car gearboxes.

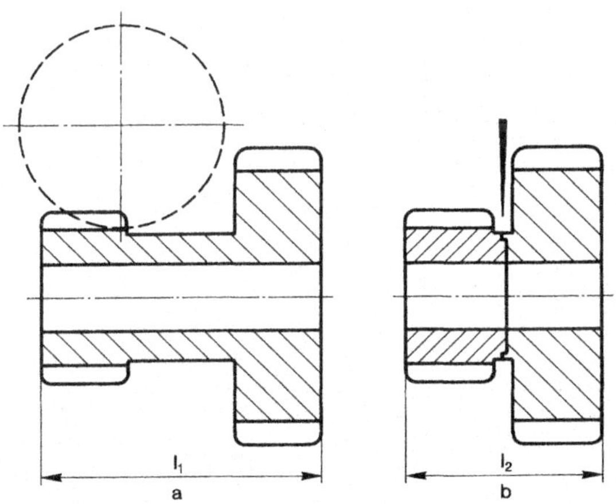

Fig. 156 Effect of the method of manufacture on the dimensions of gear wheels: *a)* Produced only by machining; *b)* Produced by machining two separate parts which are then electron beam welded together.

The generally strict requirements for dimensional accuracy in toothed gears, however, require that care be taken regarding any distortion when using circumferential electron beam welding. Figure 157 illustrates the tendency to radial distortion of axial welds as a result of transverse shrinkage at the start of welding. The workpieces should thus on no account be allowed any free play when fitted together prior to welding, in order not to increase the amount of radial distortion.

For axial welds, therefore, press fits should be chosen subject to the following limits:

- With light press fits such as H7/k6 or H7/n6 it should be remembered that with the combination of the largest hole with the smallest diameter shaft, a small amount of free play can occur.
- With excessive over-dimensions such as H7/s6, distortion again increases [125–127]. In this case the stresses generated by the pressing operation and released during welding as distortion cause increasingly greater distortion. The increased work involved in welding heavy press fits yields no additional advantages.
- The ideal press fit has been shown to be H7/r6 with H7/n6 as an upper limit.

Furthermore, it can be seen from Fig. 157 that axial welds can also show a tendency to angular distortion which must be taken into account when welding large diameter discs. This type of distortion is observed in particular in welds without roots, that is, with fusion zones having a pronounced V shape.

Also for this reason, as mentioned in section 6.3.3, workpieces including axial welds should be prepared such that a full through wall thickness weld can be made (see Fig. 128).

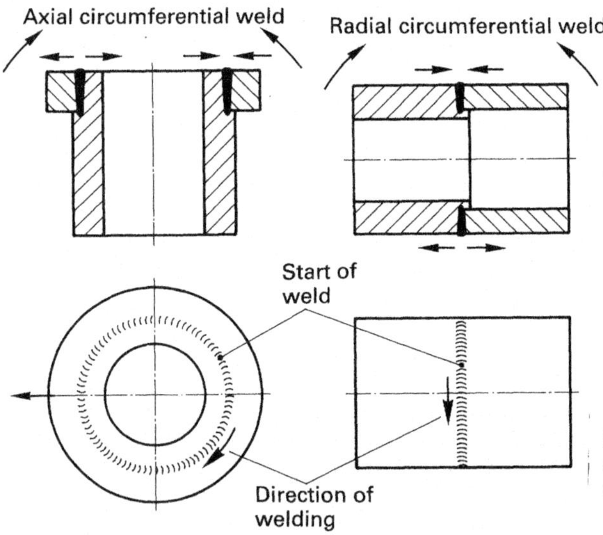

Fig. 157 The relative tendency of axial and radial welds to cause deformation on welding.

From Fig. 157 it also follows that when welding axial seams, non-uniformly distributed transverse shrinkage can lead to curvature along the longitudinal axis. This tendency can be effectively countered by use of tack welding, very stiff welding jigs, etc, or by use of wide gap free centering lips.

Distortion of circumferential welds is also dependent upon the welding parameters used, such as welding speed, weld overlap, etc. Examples of this, illustrated using simple or specific types of workpiece, are described in the literature [128–130]. The results obtained can seldom be applied to other welding tasks since the geometry of the workpiece in the region of the weld also has an important role to play. In the following examples, which are intended solely to illustrate the likelihood of distortion, the question arises as to which part of the workpiece, taking into account its function, should remain most accurately dimensioned. Should this be the outer diameter of a ring d_a, as shown in Fig. 158, which should exhibit as little radial distortion as possible, then distortion can be reduced by a design which ensures a high d_a/d_{di} ratio, or in other words a high section modulus (moment of resistance).

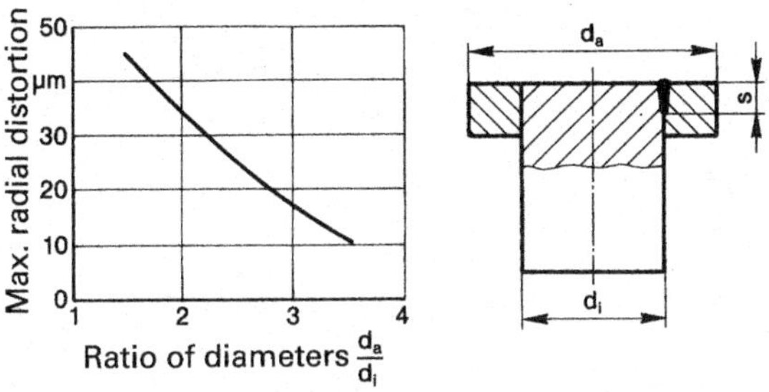

Fig. 158 Effect of component stiffness on the maximum radial distortion occurring in axial welds [125].

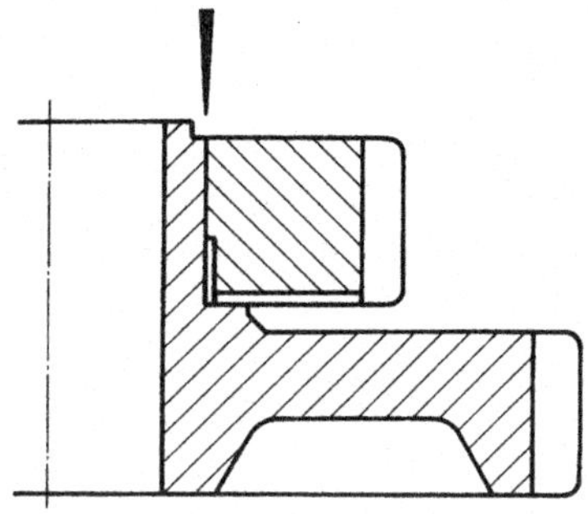

Fig. 159 Example of a gearwheel with an unsuitably located axial seam. The weld is too close to the central bore.

A mistake is often made in placing the weld too close to high accuracy reference surfaces, for example a machined bore in a gearwheel (Fig. 159). In such a case the slightest distortion on welding will impact directly on rotational accuracy in service.

It is far better either to move the weld further away from the bore (Fig. 160), or to provide a slot to take up any distortion (Fig. 161). Which of these solutions is better used to counter welding distortion can only be determined by trial welding actual components.

It cannot be determined also whether axial or radial welds are more advantageous in manufacturing. For axial seams, because of the requirement for

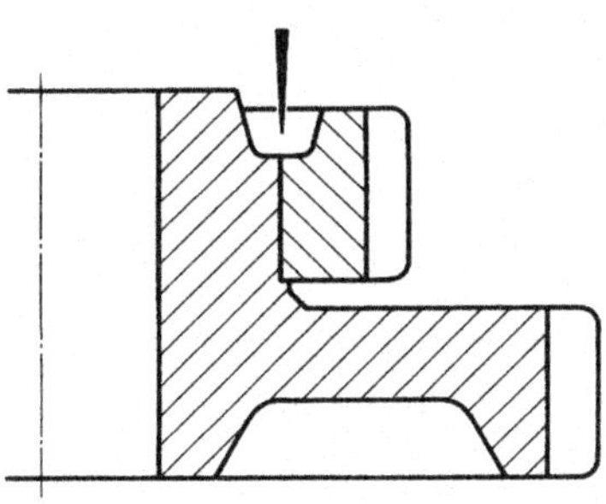

Fig. 160 A better positioned axial seam compared with Fig. 159. The weld is further from the central bore and the wall thickness has been adapted to suit the required weld penetration.

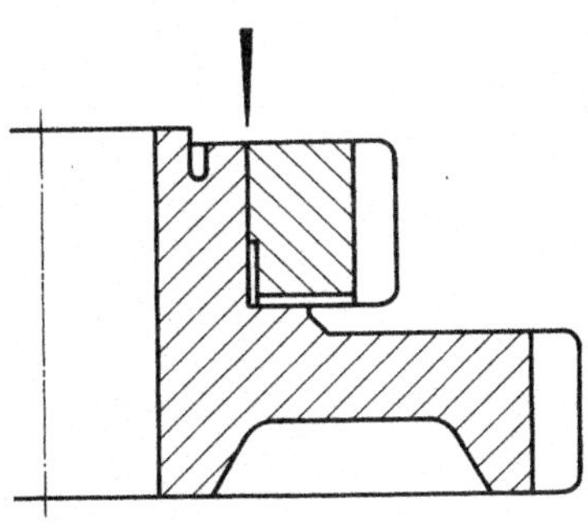

Fig. 161 A better positioned axial seam compared with Fig. 159. A slot has been included to accommodate radial shrinkage.

a press fit between components there is no additional requirement for centering or clamping jigs, and high rotational accuracies can be achieved. However, since the internal welding stresses can only be slightly reduced, with certain materials an increased tendency to cold or hot cracking must be taken into account. Workpieces with axial weld seams can also be simply and quickly set up in, and removed from, the working chamber. Thus in actual manufacturing axial welds are preferred for series welding of components.

In the case of radial welds the workpiece can shrink almost unhindered in the axial direction. However, a centering and clamping arrangement is required. Setting up and removal of a workpiece from a rotary welding head is also somewhat more involved than with axial welding because of the need for a counter clamp (see Fig. 224).

6.13 Welding assembly jigs

Jigs for electron beam welding differ only slightly from those used in other welding processes. As mentioned in section 5.5, such jigs should be made of non-magnetic metallic materials such as aluminium, austenitic steels, bronzes,

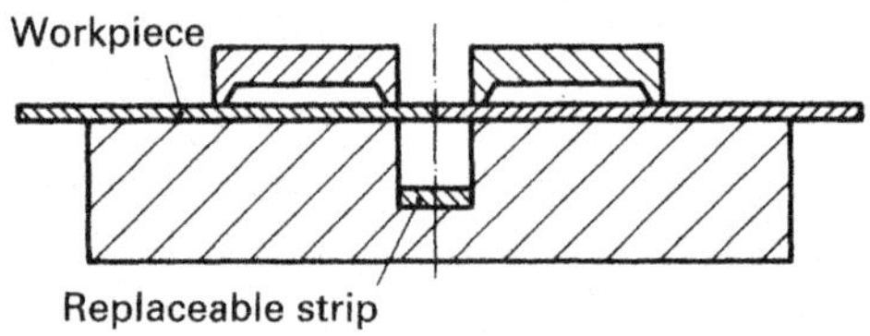

Fig. 162 Clamping arrangement for electron beam welding of sheet material. The strip beneath the root of the weld is replaceable and protects the clamping jig from being melted.

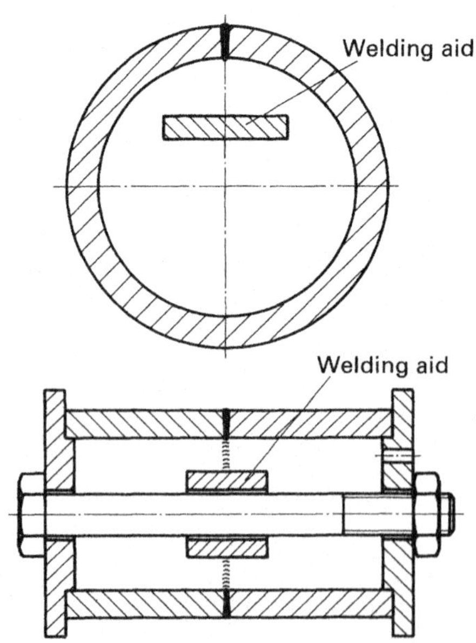

Fig. 163 An example of welding aids to protect against melting and weld spatter when electron beam welding longitudinal and circumferential seams.

etc. In the case of full penetration welds, jigs must also permit the electron beam to exit freely from the root of the weld (Fig. 162). The reason for this, as has already been mentioned, is that for the root to form perfectly and without defects the electron beam must penetrate the workpiece with a minimum excess current (the transmission current).

This residual energy can melt opposing parts of the workpiece. In practice this undesired melting is prevented by use of many and varied types of shielding (Fig. 162 and 163). In the main, plates or tube sections are used which are located at a position a suitable distance beneath the joint. Such shielding and other aids should be made from the same material as the actual workpiece because on melting, metal vapour condensate will be deposited on the free underside of the workpiece and a possible metallurgical effect must be considered. At the same time, such shields and other aids can serve as spatter

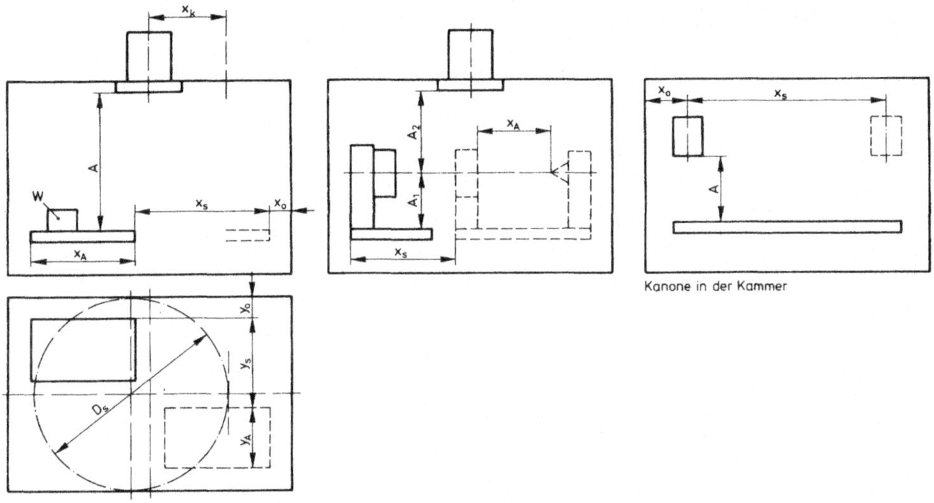

Fig. 164 The most important dimensions for working chambers, positioning equipment and location of the electron beam gun.

X_A	Length of the working/clamping table in the x-direction
X_S	Length of the welding path in the x-direction
X_O	Safety distance in the x-direction
X_K	Possible movement of the gun in the x-direction
y_A	Length of the working/clamping table in the y-direction
y_S	Length of the welding path in the y-direction
y_O	Safety distance in the y-direction
D_s;$2A_1$	Maximum diameter of circumferential seam which can be welded
A; $A_1 + A_2$	Working distance
W	Workpiece.

shields. These either catch the spatter ejected from the root of the weld or absorb so much of its energy that adherence to the welding jig or the workpiece is prevented.

6.14 Dimensions of the working chamber and moving the workpiece

In either single purpose or multi-purpose electron beam welding machines, the dimensions of the working chamber are of critical importance to the design of the workpiece and welding assembly jig which can be accommodated. It is not only the internal dimensions which must be taken into account but also, in particular the maximum possible length of weld seam which can be made in the different directions. Figure 164 shows a selection of the most important dimensions for universal working chambers equipped with either fixed or movable electron guns. The final choice is of course dependent upon the type of welding operation intended.

7 Beam and machine control

7.1 General

The electron beam differs from the heat sources used in other types of fusion welding process not only because of its high power density but also because it is ideally suited to integration into computer controlled manufacturing operations:

- The electron beam can be switched on and off at will at any position or time, with a high degree of reproducibility and without any inherent time delay. The start and finish of the weld can thus be determined on the surface of the workpiece with great accuracy. The electron beam need not, as in the case of the arc used in TIG welding, be drawn prior to starting the actual weld.
- All the parameters are electrically controllable and adjustable between wide limits. Control of an electron beam welding machine can be easily integrated, for example into a production line, in the same way as other machine tools.
- In some cases the electron beam welding machine can itself take a feedback signal from the production process and adjust the operating parameters as necessary.

The range of welding operations amongst other things also determines the beam power needed to be installed and the size of the working chamber. It also affects the requirements of the various control systems and arrangements.

The limitations of the controls of electron beam welding machines have tended to restrict their mode of operation within the total range of applications. Nevertheless in the last few years, with the multifarious possibilities presented by microprocessors, the range of use of electron beam welding has not only been considerably extended, but also improved to increase the quality of the weld produced and the ease of operation.

The following sections describe the beam and machine control systems of greatest importance to the user. On closer inspection, only the outer region of the actual welding process is thereby actually considered. For both physical and technical reasons, control of the welding process in the keyhole is possible only to a very limited degree. However, since all parameters of importance to the resulting weld are reproducible as electrical quantities to within very close tolerances, they can be used to define an optimum requirement for the stability of the welding process.

7.2 Beam current control

It has already been explained how the movement of the weld pool can lead to excess material being left at the start of the weld and craters formed at the end.

Such irregularities can be more uniformly distributed over the length of the weld by use of controlled beam current increase and decrease, to such an extent that they have almost no visible effect on the geometry of the weld. The electrical control systems necessary to achieve this were the first additional control systems to be fitted to electron beam welding machines used in manufacturing (slope control). Since then other systems have been developed for making circumferential welds which are able to control a far greater number of parameters than just the beam current.

Figure 165 shows schematically the control sequence used in welding circumferential seams. At point A on the circumference of the workpiece the electron beam welding machine is switched on to begin welding, either in displacement or time control. At point B the electron beam begins to penetrate into the material and with linearly controlled current increase the required weld penetration is finally achieved at point C. After rotating the workpiece, the weld overlaps at point A where any root cavities formed at the beginning of the weld are removed by remelting the weld metal at this point. On reaching point D, overlapping is completed and a linear decrease in the beam current begins. Up to point E, this part of the welding cycle is often the most difficult because formation of root cavities in the run out phase of the weld must be suppressed by a correct choice of settings.

For this reason, at the same time as the beam current is being decreased, the lens current I_L is increased in order to adjust the position of focus of the beam

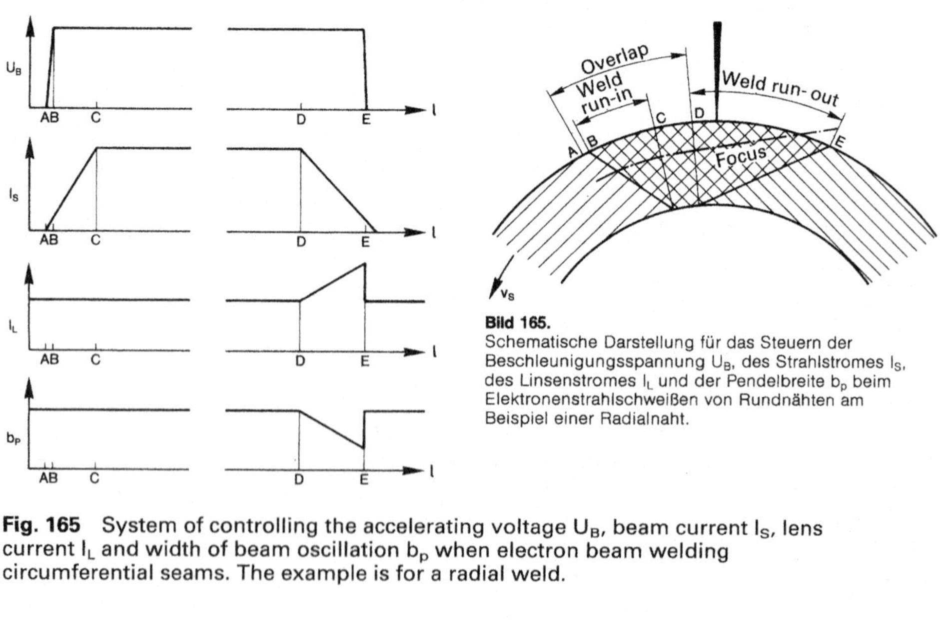

Bild 165.
Schematische Darstellung für das Steuern der Beschleunigungsspannung U_B, des Strahlstromes I_S, des Linsenstromes I_L und der Pendelbreite b_p beim Elektronenstrahlschweißen von Rundnähten am Beispiel einer Radialnaht.

Fig. 165 System of controlling the accelerating voltage U_B, beam current I_S, lens current I_L and width of beam oscillation b_p when electron beam welding circumferential seams. The example is for a radial weld.

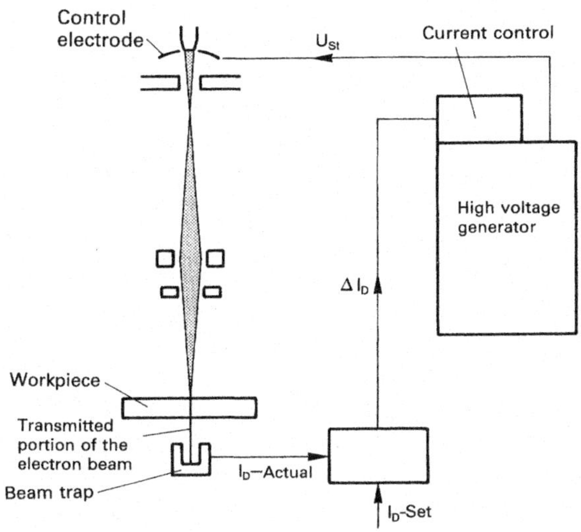

Fig. 166 A circuit for regulating the control voltage U_{st} and thus the beam current I_S by measuring the transmitted current I_D.

to suit the decreasing depth of fusion. In addition, it may be necessary to alter the beam deflection parameters, as shown for example in Fig. 165, by decreasing the amount of transverse movement b_p. During the entire welding process, both the accelerating voltage and the welding speed remain constant.

In addition to these simple beam current or parameter control systems, other systems have also been developed incorporating closed loop control. For example, in the case of adequately welded, but not excessively over welded, full penetration seams where the electron beam must emerge from the underside with a given residual energy, the transmitted current I_D may be used as a feedback signal for the control system. The principle of through thickness welding control is to measure the transmission current emerging from beneath the workpiece and use this to control one of the welding parameters in order to maintain the transmitted current as near constant as possible [131] (Fig. 166). In practice, however, the relationship between the transmitted current and the shape of the weld seam is very complex, and it is not possible to control the process universally in this way.

The main reason for this is that the effects which the electron beam has on the keyhole and the dynamics of the weld pool, which ultimately determine the size of and periodic changes in the transmitted current, are not yet fully understood. In the investigations carried out so far, it has been shown that very rapidly reacting systems are required, and that the beam current I_S, rather than beam oscillation b_p and lens current I_L, is the best control parameter to use in order to prevent instability within the control system which is otherwise prone to resonance. Furthermore, control with a horizontal beam axis is considerably better than with a vertical axis, with which the pulse frequency of the transmitted current is used as the control parameter [132]. To date, in actual manufacturing practice, these types of control system have been used only in very few cases for weld depths of up to $t = 40\,mm$.

7.3 Automatic focusing

Focusing of the electron beam by achieving maximum intensity at the point of impingement on the workpiece, although only subjective, is an accurate enough method for low beam powers. The lens current thus determined, I_L, is also valid for high beam currents as long as the beam power does not exceed a value of $P = 10\,kW$. For higher beam powers, focusing aids as discussed in section 4.4 or automatic beam focusing are required.

As there is no direct way of measuring focal position, all focusing systems evaluate certain physical effects and thus require considerable instrumentation. One such system uses the unsymmetrical current density distribution of the back scattered electrons from the keyhole and the weld pool, whilst the lens current is continuously varied between two predetermined values (Fig. 167). Once the current difference I_R between two sensors located in the plane of welding has attained a maximum, although the beam is not focused on the surface of the workpiece, the greatest potential welding penetration will have been achieved (Fig. 168) [47, 54, 56]. The threshold method (section 2.10.2) is also suitable for determining the focal position automatically. In this method the electron beam, with continuously varying lens current, is guided across the measuring sensor until the integral curve has attained the steepest gradient (Fig. 169) [57]. At this point the electron beam has achieved the highest power density and the focus will be in the plane of measurement. Using this method, the de-focusing required will still need to be determined for maximum weld penetration. Whilst automatic focusing is desirable in respect of improved reproducibility of welding, neither of these methods has been adopted in actual manufacturing practice because of the high costs and amount of instrumentation involved. Thus in practice the optimum lens current is generally determined empirically, depending upon the working distance, material, weld depth and speed of welding (section 4.4).

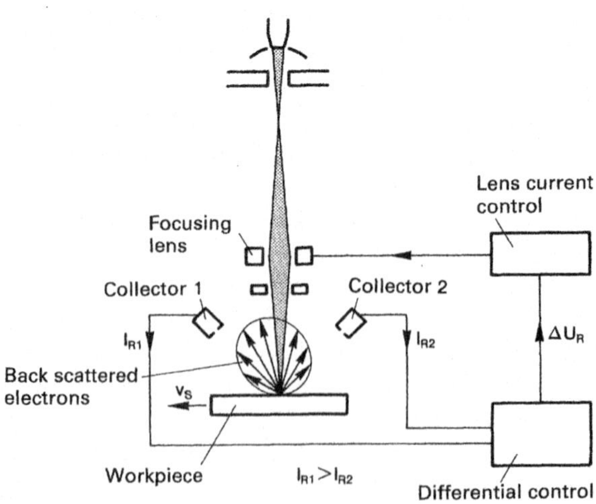

Fig. 167 A circuit for automatically focusing the electron beam by measuring different collector currents I_{R1} and I_{R2} of the back scattered electrons from the moving workpiece [54].

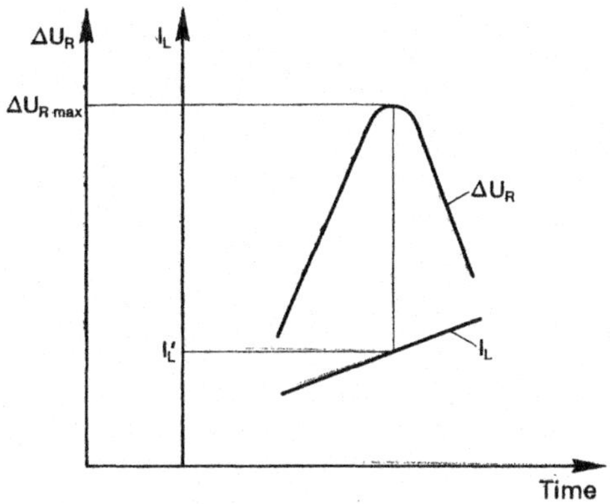

Fig. 168 Determination of the lens current I_L required to achieve the maximum depth of fusion zone s, as a function of the voltage difference ΔU_R measured from the back scattered electrons [54].

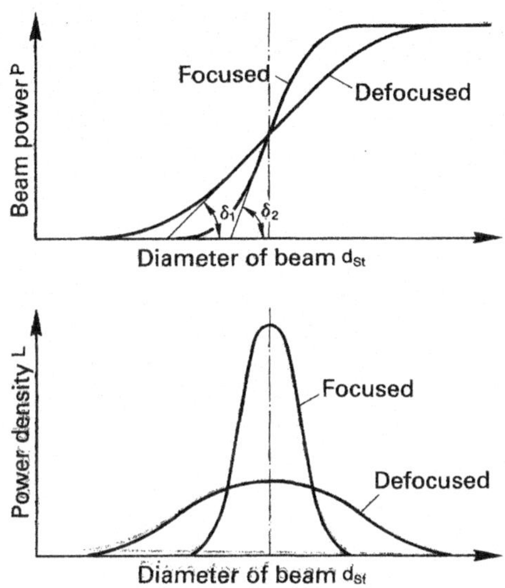

Fig. 169 Automatic focusing of the electron beam by the threshold method [57]: *a)* Integral curve of the beam power determined from measurements made using a collector (Faraday Cage). The beam focus is in the plane of measurement when the maximum slope of the curve is reached; *b)* The distribution of the power density can be calculated by mathematically differentiating the integral curve.

Another type of focusing is necessary, however, if the geometry of the workpiece requires that the lens current be varied during the welding process. This type of displacement focal control is undertaken by a computer which forms part of a comprehensive CNC system. This is discussed further in section 7.7.3.

7.4 Heating current control

In section 2.5 the technical significance of cathode heating was mentioned. In the past, after changing the cathode or after a pause in operation, the requested heating current would have been adjusted manually. Today, an automatic control system fulfills this function and at the same time takes into account both the space charge conditions in the beam source and the requirement to obtain as long a service life as possible from the cathode. Figure 170 shows the

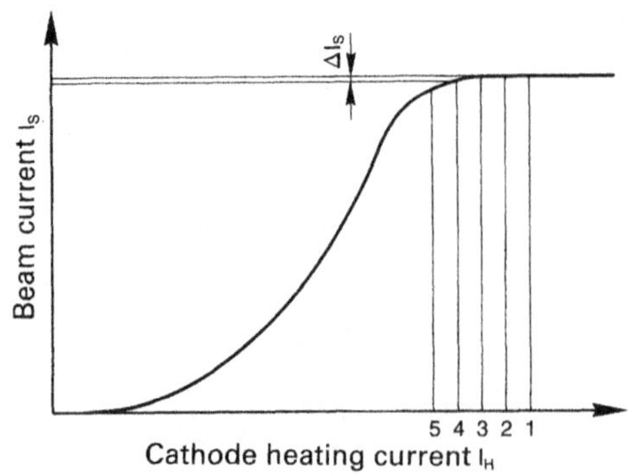

Fig. 170 Method of automatically controlling the cathode heating current.

same curve as in Fig. 11, but in this case with the beam current I_S as a function of the heating current I_H. The control initially sets the current I_{H1} and then automatically reduces the heating current in steps through I_{H2}, I_{H3}, etc, whilst the beam current I_S is measured at the same time. At the heating current value I_{H4} the control system interrupts the process because it has detected a decrease in the beam current of ΔI_S. From this it automatically sets the correct heating current I_{H2} to achieve a space charge limited condition.

If the heating current falls to a value of I_{H3}, the temperature of the cathode falls also, but assuming a constant accelerating voltage, the beam current I_S will not immediately alter. Increasing the heating current to I_{H1} would only serve to reduce the life of the cathode. Automatic setting of the heating current is thus only carried out when the cathode heating current is first applied. In this way optimum emission behaviour is ensured over the entire life of the cathode.

7.5 Seam tracking systems

7.5.1 General

Before welding can start, a butted joint must be positioned with respect to the electron beam [133]. In the simplest case this is done visually through an optical system, the workpiece and butted joint being moved to keep in line with a set of cross wires. In welding machines without optical viewing systems, positioning of the workpiece is achieved solely during the setting up procedure. If larger and heavier workpieces are to be welded, errors in positioning the workpiece can occur which can only be compensated for at great cost by use of mechanical aids. For this type of welding operation seam tracking systems have proven useful. These systems not only sense such deviations and compensate for them either by deflecting the beam or moving the workpiece accordingly, but also permit any geometrical shape of butt joint to be welded.

7.5.2 Off line

Before starting the welding sequence, a transversely scanning low power electron beam ($P \approx 30\,W$) senses the position of the butt joint along its whole

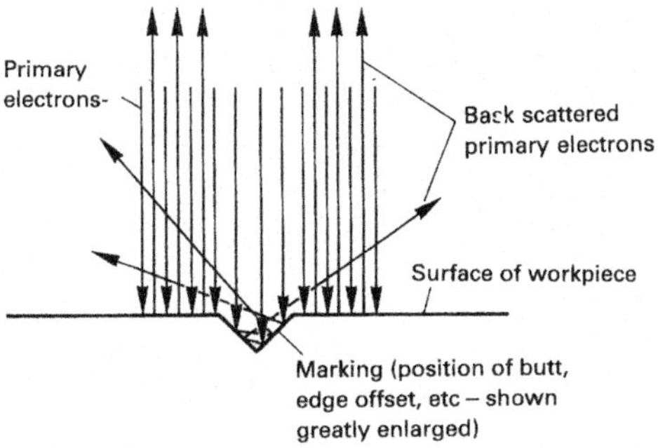

Fig. 171 Interrupted reflections of the primary electrons as the physical basis of the control parameter used for automatically sensing a butt joint using the electron beam.

length at a special rate. The electrons reflected by the surface of the workpiece have a characteristic profile (Fig. 171) which is measured by a sensor located above the position of welding and which further determines the difference in values obtained from a given specified value as a function of the position of the workpiece. Simultaneously, these data are stored within the control system. To ensure that the reflected electrons produce a clear signal, the top edges of the butt joint should be slightly chamfered or offset; a better solution, however, is to design the geometry of the workpiece such that it includes an edge which can act as a reference line. Erroneous signals reflected from surface damage, and in particular scratches, are eliminated by the control program by checking the plausibility of the values measured. On welding, the stored measurements are recalled in order to effect any correction required in the relative positions of the joint and the electron beam. In most instances a small deflection of the beam is all the adjustment that is required, but where larger adjustments are required, the position of the workpiece itself will need to be altered. The off-line joint detection system involves use of only a relatively low beam current to avoid damaging the surface of the workpiece.

The system functions well even with relatively little preparation of the joint and, as a result of the plausibility checks, is very reliable.

7.5.3 On line

If significant distortion of a butt joint is expected during welding, the electron beam itself can be used to locate the position of the joint during the welding process. In this case the electron beam accelerates ahead of the welding position for a short period of time, and rapidly senses the butt joint in the transverse direction [134, 135]. In so doing the butt joint is exposed to the full power of the beam but at such an extremely high speed that no melting of the surface of the workpiece occurs and the welding process itself is not detrimentally affected. As in the off-line system, the butt joint interrupts the regular reflection pattern of the electrons, thus providing the required measurement signal. Here, however, the signal is processed immediately during the welding process, and used to correct the position of either the beam or the workpiece.

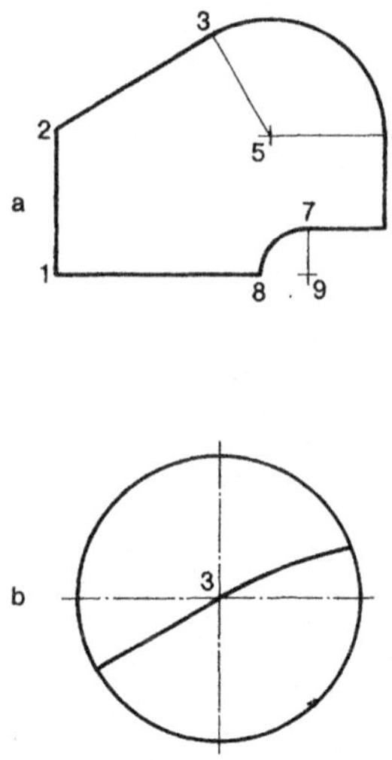

Fig. 172 Example of electron beam welding of a workpiece using a 'Teach-in' joint tracking system: *a)* Geometry of workpiece with the co-ordinate points 1 to 9 which are sequentially positioned beneath the cross wires of the eyepiece and the positions of which are then stored in the computer; *b)* An example of the positioning of co-ordinate point 3 beneath the cross wires of eyepiece.

7.5.4 Teach-in and play-back

If a specific workpiece geometry is to be welded, it may be more advantageous to store the corner points, or co-ordinates of the butt joint, prior to welding (teach-in) and subsequently to control the movement of the workpiece by computer (play-back).

To do this, each of the relevant points in the geometry is sighted beneath the cross wires in the optical viewing system, and the corresponding co-ordinates of the working table and weld positioning mechanism are stored at the press of a button (Fig. 172). In the case of curves, all that is required are the start and finish points and any other point on the curve between them or the tangent to the start of the curve. It is not necessary to store the entire profile of the joint. Once this step has been accomplished welding can begin, whereby the workpiece is moved automatically beneath the electron beam according to the co-ordinates stored. In welding series of components, the co-ordinates may be used as often as required.

The different weld guidance systems are used either individually or in combination, and significantly increase both the productivity and speed of welding machines used in manufacture.

7.6 Pump controls

Evacuation of the working chamber and electron beam gun is achieved using various types of pump which, with the aid of valves, are switched in and out according to the relative pressures achieved. This subject is discussed in greater detail in section 8.6.5. In the past it was the job of the operating personnel to switch over the valves manually at the correct time. Today, however, this task is carried out by automatic process control systems.

Pressure gauges are used to measure the instantaneous pressure during both evacuation and venting, the signals being passed to a control unit which opens and closes the valves in the system according to a preset program which ensures that each of the pumps in the system operates in its optimum pressure range. The control system incorporates a system of electronic interlocks which prevent errors in operation. The system also monitors the flow of cooling water to the pumps, the compressed air, the power supply to the servo-valves and the number of hours of operation until the next service. Warnings of failures in any parts of the system are signalled by both acoustic and optical indicators. In the case of production line welding machines, the pumping system forms part of the overall process control and control of the vacuum system is integrated into the welding cycle.

7.7 Control systems

7.7.1 General

On earlier types of machine, control systems generally only governed the reduction in beam current when welding circumferential seams, the oscillation of the beam and the switching of the vacuum servo-valves. These integrated control systems incorporated protective cut-outs, relays and simple electronic elements, the overall operation of the system being determined by the wiring, and being modified only by altering the hardware itself.

With today's microprocessor control systems, however, there is now no need to modify the internal electronics if changes in operation require to be made. This is now achieved much more easily by modifying the software program. The numerous programs which can be stored in such semi-conductor based systems thus permit many different types of operation to be carried out using one and the same system of electronics. Today, these types of system are simply referred to as programmable logical control (PLC) systems.

If more complex control functions need to be carried out, for example when welding using seam tracking systems to produce joints in three dimensions, then computers are used which determine and carry out the individual operations fully automatically according to a pre-programmed sequence. These types of system are known as computer numerical control (CNC) systems.

Today's modern control systems offer many different possibilities in beam guidance and release the operating personnel from much routine work in operating the machine. In addition, with use of additional monitoring arrangements, they also increase the overall reproducibility and quality of the welds produced.

7.7.2 Programmable logic control systems

These process control systems incorporate very powerful control modules which can control the individual parameters and functions (also called axes) of the process according to a predetermined program sequence (Fig. 173). In electron

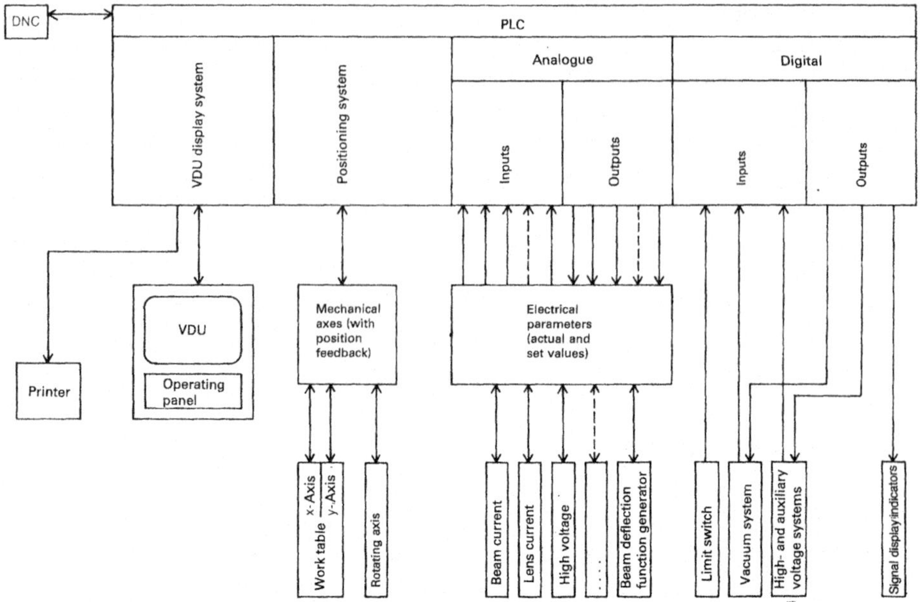

Fig. 173 An electron beam welding machine with a PLC system [136].

beam welding machines one differentiates between the electrical parameters such as

Beam current	I_S
Lens current	I_L
Deflection current	I_K
Oscillation frequency	f_P etc,

and mechanical parameters such as

Movements in the X direction
Movements along the A axis, etc.

Programmable logical control systems are employed for the following operations:

In the operation of the welding machine:

- Monitoring the mechanical, electrical and vacuum systems;
- For warning of malfunctions and fault diagnosis;
- For indicating servicing requirements.

For the workpiece:

- To determine the position of the workpiece;
- To control the movement of the workpiece.

During welding:

- To enter the welding parameters (the set parameters);
- To enter welding programs, where the parameters are controlled independently of one another;
- To give commands to the mechanical, electrical and vacuum systems;
- To determine the actual welding parameters (the measured parameters);
- To compare the welding parameters set with those actually measured and to indicate differences;
- To store and document the parameters;
- To recall the welding parameters set for repeated use.

Visual display purposes:

- To provide VDU process information in clear text and with diagrams;
- To enable the operator to interact through dialogue to control the welding machine;
- To indicate the operating condition of the machine, to enter commands and to receive an indication that these have been carried out.

These possible methods of control have had an important effect on the overall economic viability of the electron beam welding process.

Examples of this are:

- Automatic control of the vacuum system
  Pressure dependent switching of the valves and pumps can be controlled, see section 8.6.5, taking into account minimum evacuation times for the working chamber and if necessary for the electron beam gun after the cathode has been changed, and preventing incorrect operation. The state of the vacuum system together with information regarding the number of hours to the next service, oil change, etc, is also indicated in clear and easy to understand diagrams.
- Optimisation of cathode heating
  After the cathode has been replaced it is heated automatically, see section 7.5, and the heating current applied during this process optimised. In addition to a series of electron optical advantages, this also helps to achieve the maximum service life from the cathode.
- Faster determination of the welding parameters to be set
  Previously, the optimum beam current I_S needed to be determined by trial welds, with a separate weld being carried out for each different setting. If the welds were carried out next to one another, the results obtained were of real practical use only if the workpiece were allowed to cool after each weld run.

  A much shorter and more efficient way, however, is to make a single trial weld with a successively increasing beam current.
- Control of the welding parameters set

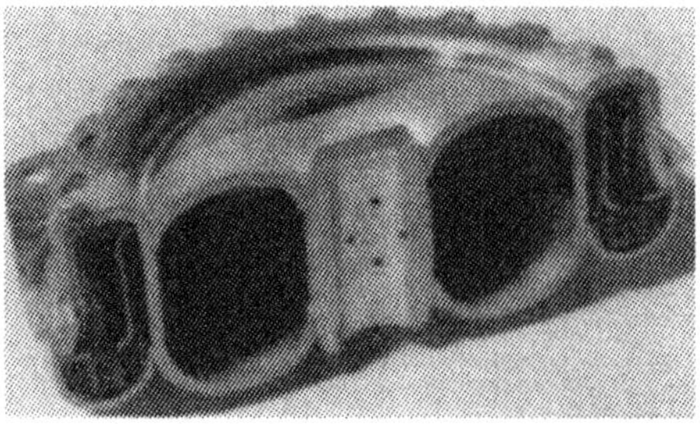

Fig. 174 Five axial circumferential welds join parts pressed from AlMg 5 sheet to form a housing for an airbag safety system.

The actual parameters are measured during the welding process and compared with the set values:

- Accelerating voltage U_B
- Beam current I_S
- Lens current I_L
- Welding speed V_S
- Deflection current I_K
- Beam deflection or oscillating current I_K; I_p
- Working pressure p_A

If the limits set for the values of any of these parameters for the particular welding operation are exceeded, an optical or acoustic warning signal is given.

- Welding of components
  The parameters used for welding a particular component can be stored under a particular identification code and can be called up and re-entered into the machine control system. In this way, welding of series of components can also be repeated in any desired sequence.

As a practical example, Fig. 174 shows part of an airbag system designed to protect the head and upper body of a driver in a collision [137]. Five joints have to be welded with the gas charge already in the casing, having fusion zone depths of between 2.5 and 4 mm. The welding parameters set for the respective joints, and in particular the beam current I_S, the lens current I_L, the welding speed V_S and the diameter of the circumferential seams are stored in the control system and are called up automatically according to the program and monitored during the welding process to ensure that the very close tolerance limits set are not exceeded.

PLC systems with their numerous applications have now become a central component in modern electron beam welding machines. They are able to process both analogue and digital values and can control axes independently of one another using various different modules. This type of control is, however, independent of both the distance welded and of time.

7.7.3 Computer numerical control (CNC)

Programmable logic control systems are characterised by the logical interaction which takes place between the programmed command given and its execution.

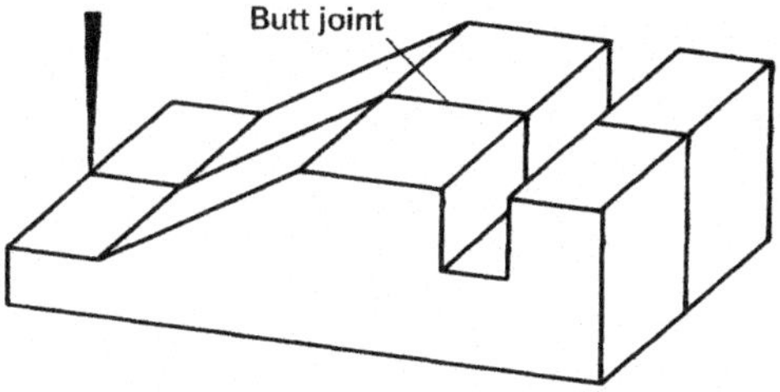

Fig. 175 Computer numerical control of the welding and lens currents as a function of the weld position.

They are thus particularly suitable for storing fixed welding sequences made up of variable parameters in the machine's control system.

If, however, numerous very different welding sequences need to be carried out, requiring very flexible and mutually interactive control of a large number of parameters, then this task is best performed by a computer numerical control system. As an example, Fig. 175 shows a workpiece which requires welding with position dependent beam and lens current control. The advantages of CNC systems are their inherently rapid processing capability, the ease with which they can be used to set up very complex welding program sequences, and their ability to link and interpolate the various parameters. Computer numerical control systems thus provide a very flexible link between the movements of the workpiece and the subsequently dependent welding parameters. These types of control system thus present a useful supplement to PLC systems (Fig. 176).

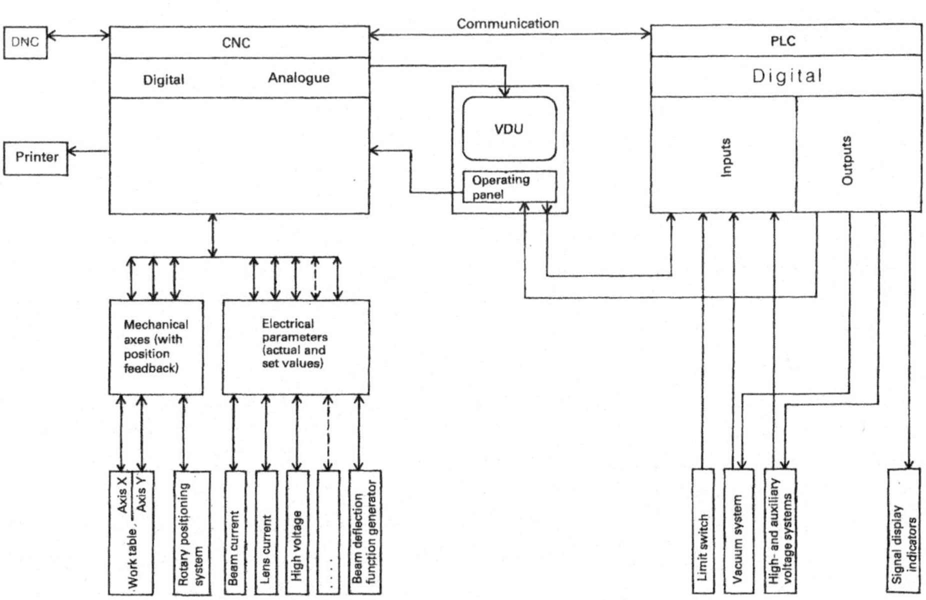

Fig. 176 Block diagram of an electron beam welding machine with a PLC and CNC system [136].

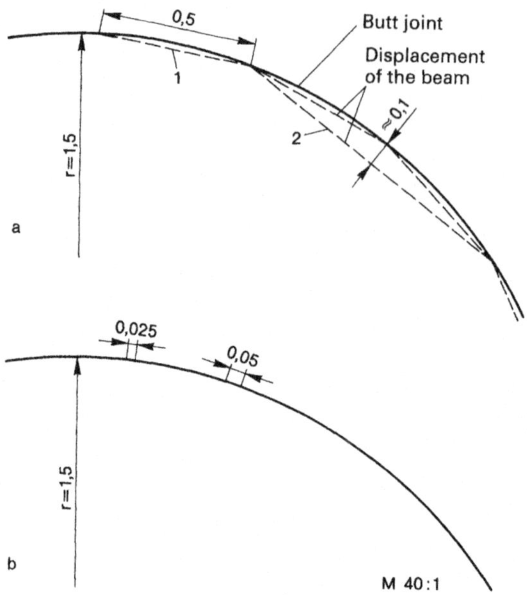

Fig. 177 Deviation of the beam spot from the butt joint when welding with different positional feedback update rates.

CNC systems have already been used for a long time with machine tools and are now also installed on electron beam welding machines. It has, however, been clearly demonstrated that when using all the many possibilities of electron beam welding, the machine control systems currently commercially available can only partially fulfill the requirements of industry. For some welding operations they are proving too slow and in many instances are unable to control a sufficient number of axes independently of one another.

Powerful CNC systems are characterised by having update rates as high as possible. This is the rate at which the program commands are passed on to the axes by the computer, and at which the actual and set parameter values are compared with one another and if necessary adjusted. For electron beam welding machines the update rate should be <2 ms (current achievable rate: 0.1 ms). In contrast, the update rate for machine tool CNC systems is considerably higher. Another measure of the power of such systems is their data reading cycle time. This expresses the time required by the computer to read a command. If this is greater than the update rate then on reading a fresh command the computer will have to wait through one or more cycle times until the command can be passed on to the relevant axes. In CNC systems specially developed for electron beam welding machines, the read cycle time is of the same order of magnitude as the update rate, or is reduced almost to zero by use of specially developed programs.

The following example illustrates the effect which the update cycle time and read rates have on the accuracy of guidance of the electron beam, or workpiece. An arc of radius 1.5 mm is to be welded, an example which might occur where two weld seams at right angles to one another require to be joined in a radius. Figure 177 shows a section of the arc increased to a scale of 40:1.

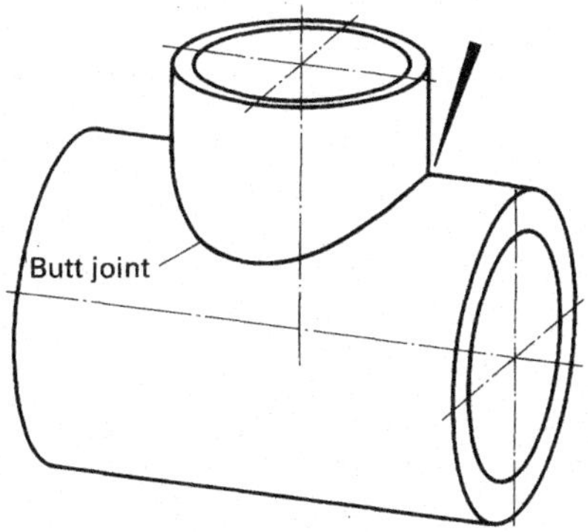

Fig. 178 A workpiece suitable for welding with computer numerical control of a number of the electrical and mechanical axes.

At an assumed welding speed of $v_S = 25\,mm \cdot s^{-1}$ and with an update rate of 20 ms, the electron beam will travel a distance of about 0.5 mm in a straight line before it receives its next guidance command. This is shown by the dashed line 1 in Fig. 177. In doing so the beam describes a polygon and deviates significantly from the required line of the arc. At a welding speed of $v_S = 50\,mm \cdot s^{-1}$ this deviation increases to about 0.1 mm, shown by the dashed line 2 in Fig. 177, which in the case of high quality welds can lead to the occurrence of significant weld defects. Read rates which are greater than the update rate and thus necessitate co-ordinate points being missed out increase the error still further. Here, it must be taken into account that the overall read cycle time also depends upon the number of axes to be controlled. The read cycle time can thus add up to a quite considerable timespan if, as has already been mentioned, this is not reduced by special programming methods.

In comparison with an update rate of 1 ms and the same welding speeds mentioned above, the electron beam travels straight line distances of about 0.025 and 0.05 mm respectively. In this case, however, in spite of an increased scale of the drawing, the polygon travelled by the beam or workpiece cannot be differentiated from the line of the arc (Fig. 177*b*).

It can thus be seen that in electron beam welding, only use of a very high update rate can ensure that deviation in the position of the beam from the joint, in particular when welding curved seams or with many varying parameters, is kept to an absolute minimum (Fig. 178).

In comparison with PLC systems, the particular ability of CNC systems to control a number of parameters and machine functions simultaneously and independently, and to interpolate between them, opens a wide range of additional and interesting possible applications for electron beam welding in both development and manufacturing. The following summary lists the current range of controllable axes which might be found on a modern machine used for electron beam welding of large turned components:

- Accelerating voltage U_B
- Beam current I_S
- Lens current I_L

–	Pulse current	I_{SP}
–	Pulse frequency	i_f
–	Pulse time ratio	i_V
–	Beam deflection in the X direction	I_K
–	Beam deflection in the Y direction	I_K
–	Beam oscillation in the X direction	I_P
–	Beam oscillation in the Y direction	I_P
–	Beam oscillation vectoring in the X direction	–
–	Beam oscillation vectoring in the Y direction	–
–	Constancy of the width of beam oscillation b_p at different working distances A_A	–
–	Movement of the workpiece in the X direction	V_X
–	Movement of the workpiece in the Y direction	V_Y
–	Movement of the workpiece in the Z direction	V_Z
–	Movement of the workpiece about the A axis	–
–	Movement of the workpiece about the B axis	–
–	Movement of the workpiece about the C axis	–
–	Positioning of the optical viewing mirror	–
–	Aperture of the optical viewing system	–
–	Movement of the electron beam gun	–
–	Working pressure	p_A

The following illustrates a number of practical examples of electron beam welding with CNC systems.

Figure 179 shows the problem often encountered of having to weld a seam the longitudinal axis of which does not coincide with any of the axes of movement of the working table. Initially the co-ordinates of the start A and end B of the butt joint are determined and stored by positioning them beneath the cross wires of the optical viewing system (see section 7.5.4). At the start of welding the computer guides the working table simultaneously in both the X and the Y directions enabling the electron beam to follow the angle of the butt joint.

The set values of the co-ordinates are fed to the positioning mechanism of the working table, for example every 1 ms. A similar process is used in welding axial circumferential joints where deviations between circumferential weld seam and the centre of rotation occur (Fig. 180*a*). After entering the co-ordinates M_1 and M_2 of the centre point and the diameter of the circumferential seam d_R and depending upon the construction of the welding machine, the computer guides the joint beneath the axis of the beam, or deflects the electron beam to the position of the joint. If, however, the roundness of the weld seam is not guaranteed, in the main because of distortion occurring on welding, then an on-line seam tracking system must take over control of the positioning of the beam (Fig. 180*b*). The same general method is used with radial circumferential welds (Fig. 181), where here the co-ordinates of the axis of rotation are programmed. If, as a result of eccentricity, variations in the working distance A_A occur in axial and radial seams, then once programmed the CNC

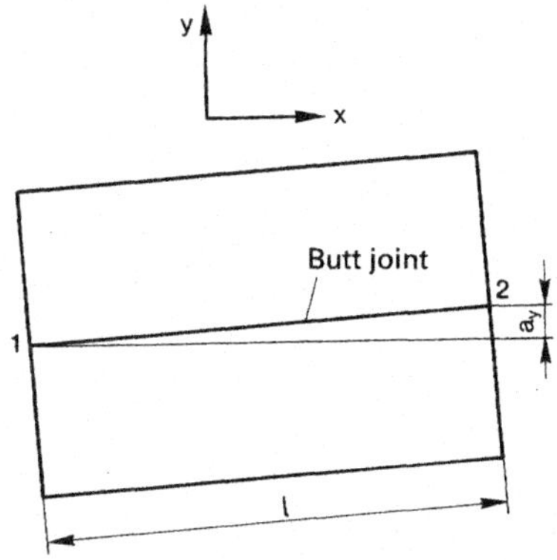

Fig. 179 A butt joint the longitudinal axis of which is not coincident with the co-ordinate axes of the positioning system (work table).

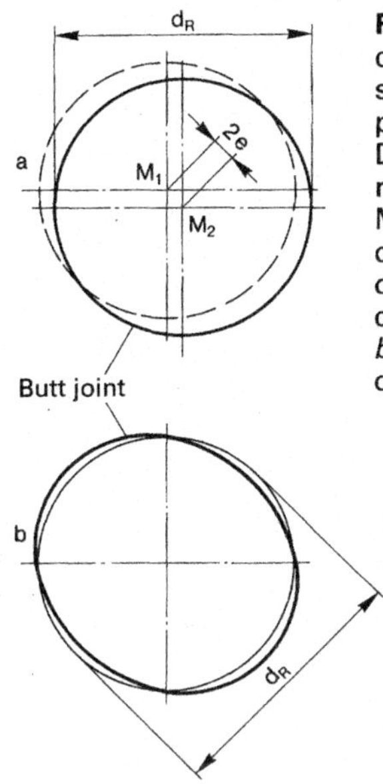

Fig. 180 Axial circumferential butt weld showing various types of positional deviation: *a)* Deviation from the axis of rotation, e = Eccentricity; M = Centre point of circumferential weld; d_R = Diameter of the circumferential weld; *b)* Deviation from the line of the circle.

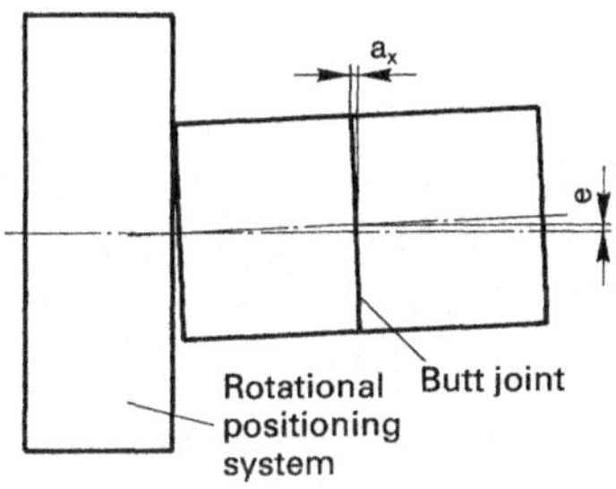

Fig. 181 A circumferential butt joint with deviation from the axis of rotation.

system will also take over the task of varying the lens current I_L and thus the position of the focus relative to the rotation of the workpiece.

Figure 182 shows a cross section of a turbine blade to be welded.

The number of sections into which the surface to be welded is divided in order to adapt the welding parameters (e.g. beam current I_S, lens current I_L, welding speed V_S, beam oscillation, etc) to the changing working distance A_A and material thicknesses t, depends upon the actual dimensions of the blade as well as on other factors such as the material type. If the most exact beam guidance possible is required, then the co-ordinates of the different arc segments of the profile can be entered into the control system. By continual interpolation, the computer then takes over complete control of the welding parameters. The run-in and run-out areas needed to prevent defects at the beginning and end of the weld are not shown in Fig. 182.

The function generator needed for oscillating the beam is also controlled by the CNC system. If axial circumferential seams are to be welded with transverse beam oscillation then the direction of oscillation in relation to the centre point of the circumferential joint is vectored (Fig. 183). In addition, the oscillation amplitude b_p is also controlled as a function of the lens current I_L

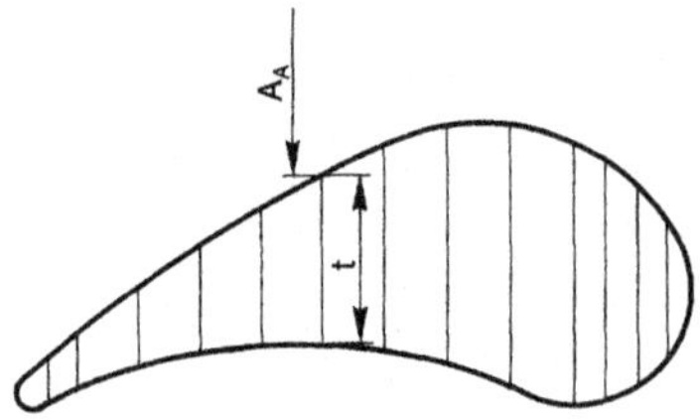

Fig. 182 Cross section of a turbine blade for welding with varying working distance A_A and with varying wall thickness t.

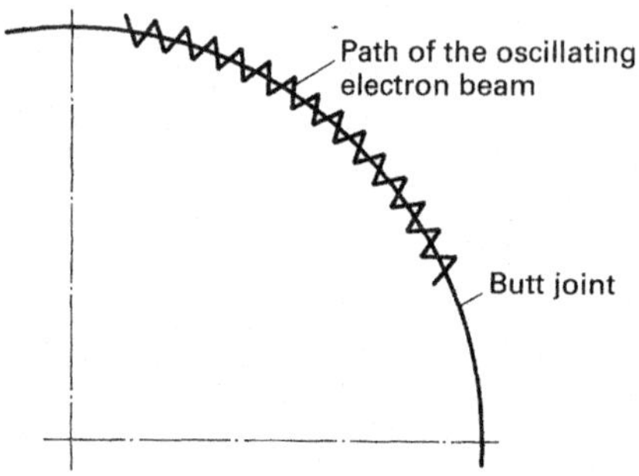

Fig. 183 An axial circumferential butt weld to be made with beam oscillation. The direction of oscillating is towards the centre of the circumferential seam.

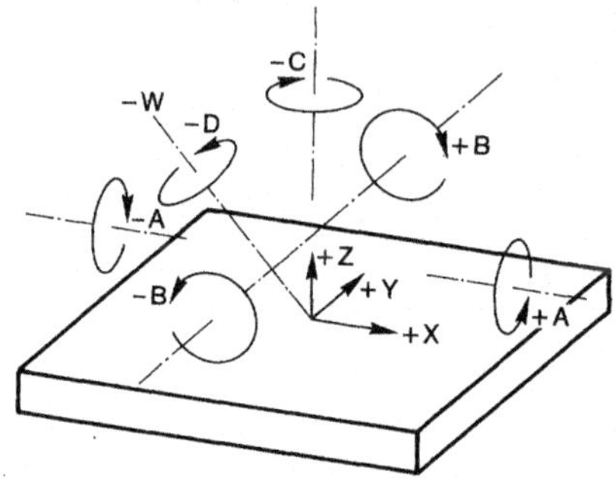

Fig. 184 Identification of the axial co-ordinates and directions of rotation used in the computer numerical control systems of electron beam welding machines, and as in DIN 66 217.

and the working distance A_A. In other words, by altering the position of focus or the working distance, the oscillation amplitude at the surface of the workpiece can be maintained constant.

The identification of the axial co-ordinates and directions of rotation used for CNC systems are specified in DIN 66217 (see Chapter 12).

A right handed, rectilinear system has been adopted with the axes X, Y and Z being coincident with the principal direction of movement of the welding machine and the workpiece set up for welding.

If, as in most electron beam welding, it is the workpiece which is moved, then the directions of movement and of welding will oppose one another. Thus, to weld in the +X direction, the workpiece must be moved in the −X direction. If, however, the electron beam is moved, then both directions of movement will be coincident (Fig. 184). The letters A, B and C denote rotational movements, the rotational axes of which are parallel to the axes X, Y and Z respectively. The direction of rotation is considered to be positive when the rotational movement is clockwise when viewed in the positive direction of the co-ordinate axis. An axis of rotation not coincident with one of these principal co-ordinates is denoted by the letter W, and a rotation about it by the letter D.

8 Electron beam welding machines and equipment

8.1 Basic types of construction

With the exception of a few special cases, electron beam welding machines can be divided into three groups:

- Universal welding machines;
- Cycle type production machines;
- Continuous welding machines.

All these types of machine include the following common elements:

- An electron beam gun;
- A working chamber;
- A method of positioning the workpiece;
- A high voltage supply;
- Control and adjustment instrumentation;
- A vacuum system;
- Welding jigs.

These elements are arranged around the working chamber which is the central feature of every electron beam welding machine (Fig. 185). In addition there is a control desk which contains all the most important operating elements of the system together with any data entry or monitoring instrumentation and which may be supplemented by a VDU in the case of a PLC or CNC system (Fig. 186). The vacuum pumps are situated directly behind the rear wall of the chamber and are acoustically insulated.

The electron beam gun is connected to the high voltage power supply by a special cable, and depending upon the welding operation and arrangement of the working chamber can also be moved and operated in the horizontal position. Universal welding machines are generally constructed with a working table which can be moved out of the working chamber to facilitate setting up of the workpiece in a rotary weld positioning mechanism or other welding jig (Fig. 187). Cycle type and continuous welding machines have working chambers which are specially designed to suit the particular geometry of the items to be welded, and are integrated into the production line for the particular product.

8.2 Electron beam guns

Figure 188 shows schematically the construction of an electron beam gun. It is made up of an upper chamber in which the electron beam is generated

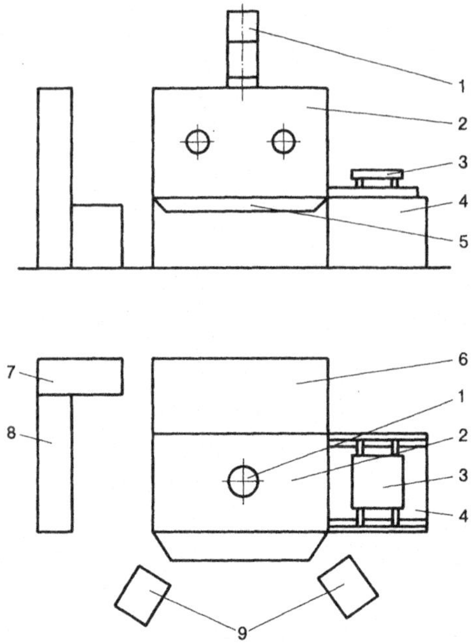

Fig. 185 The components of an electron beam welding machine:
1 Electron beam gun
2 Working chamber
3 Work table which can be moved out of the working chamber
4 Supporting bed
5 Covers of the vacuum system
6 Vacuum pumps
7 High voltage power supply
8 Monitoring and control cabinets
9 Operating desk and VDUs for the PLC and CNC systems (Not shown in the front view).

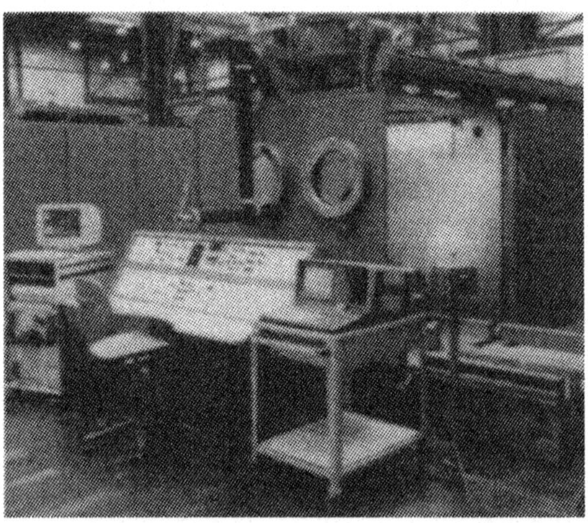

Fig. 186 Front view of an electron beam welding machine with the working chamber open. $V = 3\,m^3$. In the foreground, from left to right: VDU, printer, and computer for the CNC system, operating desk, VDU for the PLC system.

and which contains the high voltage connection, the insulation casing and the triode system, and a lower chamber containing the optical viewing system, the focusing lenses and the deflection system. The two areas of the gun are also separate and isolated parts of the vacuum system. Separation of the two areas by the electron beam guns' column valve makes it possible to maintain the vacuum in the beam generator whilst changing the workpiece, i.e. with the working chamber open and at atmospheric pressure.

If the working chamber has been evacuated to begin welding, the col-

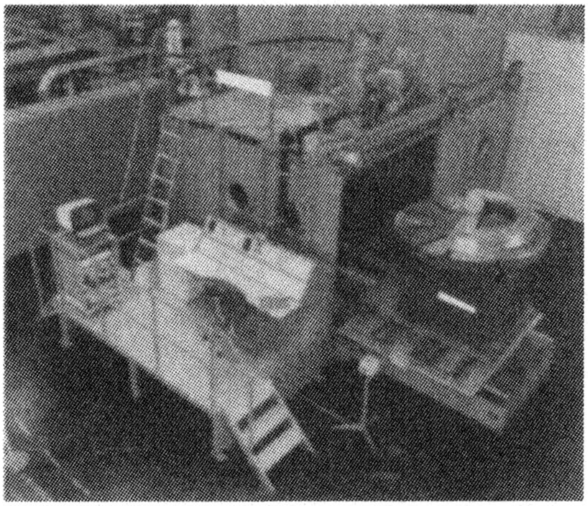

Fig. 187 Front view of an electron beam welding machine with the working table run out of the working chamber showing the rotary positioning arrangement. Maximum beam power P = 30 kW, v = 10 m³.

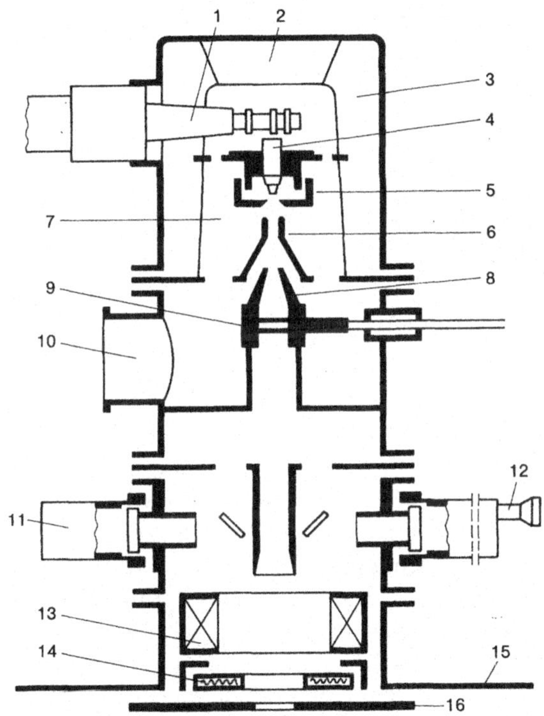

Fig. 188 Cross section of an electron beam gun:
1 High voltage cable
2 Lid of electron gun which can be opened to change the cathode
3 High voltage insulator
4 Cathode exchange cartridge
5 Control electrode
6 Anode
7 Beam generator space
8 Pressure stage
9 Column valve
10 Connections for evacuating the beam generator
11 Lighting for the optical viewing system
12 Viewing system with telescope
13 Focusing lens
14 Beam deflection system
15 Part of the working chamber
16 Heat shield.

umn valve must be opened before turning on the electron beam. This valve is generally connected to a pressure stage which, together with a special pumping arrangement (Fig. 189), is able to maintain a pressure difference between the working chamber and the beam generator of $p_A = 10^{-2}$ and $p_E = 5 \cdot 10^{-5}$ mbar respectively.

The top of the beam generator can be opened, when flooded, and the heated

Fig. 189 Electron beam gun with part of the vacuum pumping system (left) for evacuating the beam generator.

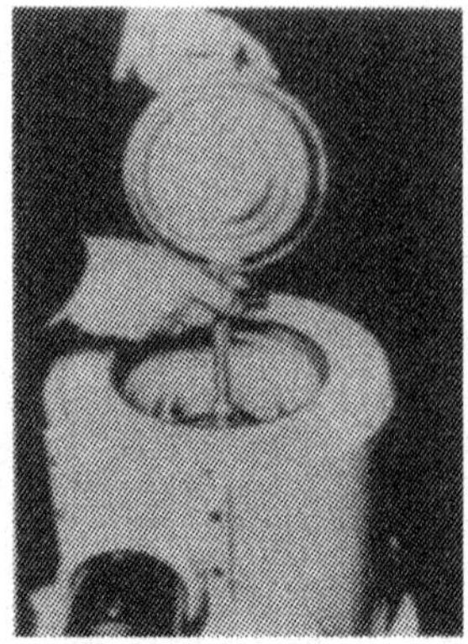

Fig. 190 Removing the cathode exchange cartridge through the top of the open electron beam gun.

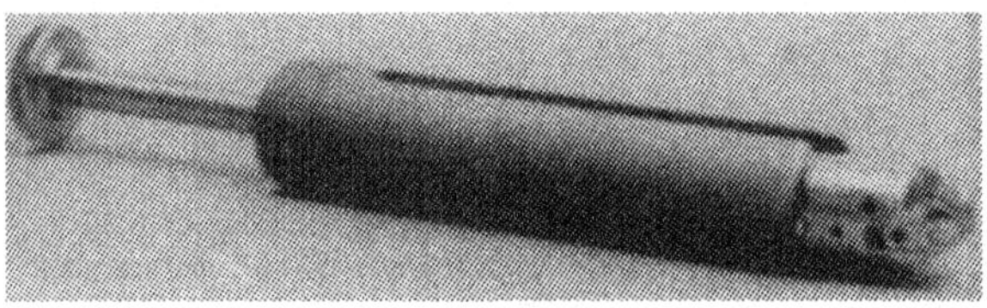

Fig. 191 Tool with cathode exchange cartridge.

ribbon carrier removed using a special tool to permit the cathode ribbon to be replaced (Fig. 190 to 193). With the aid of a loading jig, a new preformed tungsten ribbon to the correct size can then be clamped into position (Fig. 192). This type of clamp ensures perfect contact for the heating current and that the position of the emitting surface remains constant even at the high temperatures ($T > 2500\,°C$) reached by the cathode. Without altering the very accurate, close tolerance position and the axial alignment of the electrodes, the heated ribbon carrier is replaced into its original position in the beam generator and the lid closed before evacuation. In this way a very high degree of reproducibility is ensured in the beam generator after changing the cathode.

Mainly in universal welding machines, the lower part of the electron beam gun also includes an optical viewing system for positioning a butt joint, determining the lens current during focusing and for inspecting the welding spot (Fig. 194). This generally consists of a telescope with binocular eyepieces (Fig. 195) which permits accurate positioning of the electron beam at the required point of welding. A source of light is also provided to illuminate the area of welding, although in some instances sufficient light is provided by the

Fig. 192 Jig for changing the filament.

Fig. 193 Cathode exchange cartridge with filament in position.

Fig. 194 Positioning the workpiece whilst evacuating a universal electron beam welding machine.

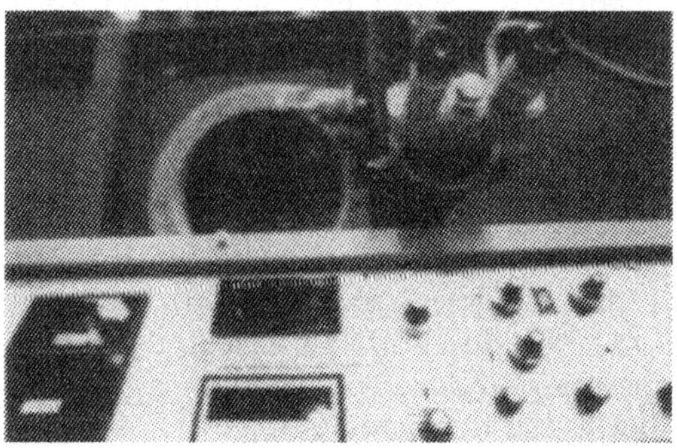

Fig. 195 Operating desk and optical system.

glow from the heated cathode to illuminate the surface of the workpiece for positioning the point of welding. During the welding process, protective glasses shield the optical viewing system from the metal vapour and spatter produced, whilst a heat shield protects the focusing lens and beam deflection system from thermal damage.

8.3 Working chamber

The evacuated working chamber must not only be capable of accepting one or more of the workpieces together with any welding jigs or workpiece positioning equipment, it must also meet a number of stringent stability requirements. In particular, the air pressure which acts on an evacuated working chamber must be taken into account in the construction.

It is not sufficient just to design the chamber to have adequate mechanical strength, since the permissible distortion which may occur must also be taken into account in its construction. Such distortion affects the beam axis of the electron beam gun where it is joined on to the chamber, the position of the workpieces in the welding or clamping jigs, and the accuracy of the movement and deflection systems. Large working chambers, in particular when they deviate from cubic or cylindrical shapes as is often the case in manufacturing machines, are thus frequently designed with the aid of finite element analysis in order to minimise distortion.

The working chamber, like the outer parts of the electron beam gun, is constructed from normal commercially available structural steel which considerably reduces the interfering effects of external magnetic fields. The internal walls do not require any particular protection against rusting as the working chamber is evacuated for most of the time. In addition to this, when in operation a layer of metal condensate is continuously deposited on the surface of the internal walls of the chamber. If particularly demanding requirements are placed on the working chamber in terms of cleanliness, the internal walls can be clad with metal plates which can then be easily removed for cleaning. In fitting such plates an adequate distance must be left between the plates and the wall of the chamber in order not to unduly lengthen the evacuation process (see section 8.6.6).

Depending upon its size the working chamber is equipped with one or more windows to allow the inside of the chamber to be viewed. To protect the operator against the effects of X-rays, the windows are glazed with a special lead containing glass (see section 11.4). The doors of the chamber are also electronically interlocked with the operating cycle of the vacuum system to prevent or permit access to the chamber at the appropriate time.

The size of the working chamber depends upon the dimensions of the workpieces to be welded. Cycle type machines for large numbers of small articles are equipped with working chambers only a few litres in volume which can be evacuated in only a few seconds. For larger workpieces, working chambers in excess of $160\,m^3$ have been built [138]. Today, the machine

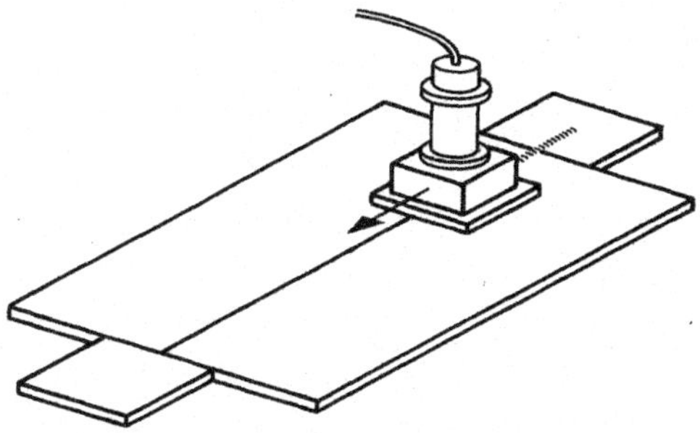

Fig. 196 An electron beam gun equipped with a mobile vacuum system [139].

manufacturers generally offer ranges of standard working chambers increasing in volume from about 0.5 to 40 m^3 (see section 8.7).

For electron beam welding with mobile and localised vacuum systems, a special type of working chamber is used. This type of system is used in particular for welding sheet or plate material, containers and pipe walls of any desired size, and here the workpiece itself forms part of the working chamber (Fig. 196).

The dish shaped working chamber containing the electron gun slides across the surface of the workpiece to be welded, and is equipped with a number of special pressure stages in order to be able to maintain the vacuum at the point of welding. Electron beam welding machines of this type have already successfully proven their viability in the laboratory [139–141].

8.4 Positioning equipment

This equipment has the task of moving the workpiece and electron beam relative to one another. In a few cases the electron gun is moved along the workpiece inside the working chamber. In general the electron gun is situated outside the working chamber in a stationary position, and it is the workpiece which is moved at the chosen speed. In some types of manufacturing machines the electron gun can be moved across the wall of the chamber, for example to be able to move the axis of the beam to the respective positions when welding axial seams of different diameters.

The two most commonly used pieces of equipment are a traversable working table and a rotary table. For welding longitudinal seams the traversable table moves the workpiece in both the X- and Y-directions, and more seldom also in the Z-direction (Fig. 197, left). The table contains T slots to enable the workpiece to be securely clamped in position, and is powered by special motors capable of operating under vacuum.

The table is driven from the motor through a toothed rack or by a compensating ball screw arrangement. Incremental displacement transducers are fixed to the table to signal the actual position of the table at any point in time, and if this differs from the required position it can be corrected by the machine and beam control systems already mentioned in section 7.5 (Fig. 198). The running and guidance rails of the table in particular are protected against

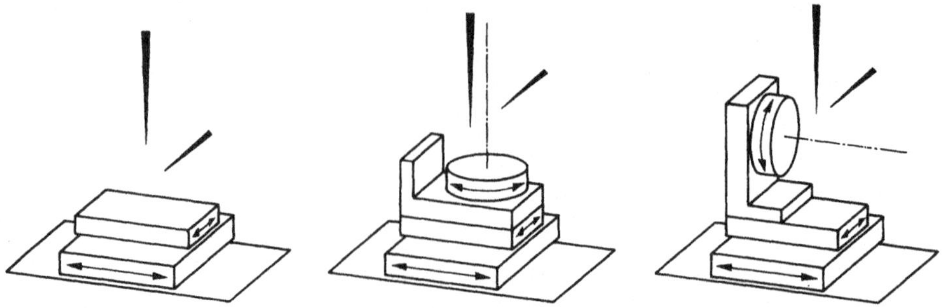

Fig. 197 Various arrangements for positioning and moving the workpiece for fixed vertical or horizontal electron beam guns: Left: Work table; Centre: Rotary positioning system with C-axis; Right: Rotary positioning system with A-axis.

Fig. 198 Incremental measuring transducer encoder for determining the actual position of the work table (encapsulated in a casing and fixed to the shaft of the larger of the toothed belt drive gears).

Fig. 199 Work table with a carrying capacity of 1.2 t on its supporting bed.

dirt, spatter and metal condensate (Fig. 199). With ball screw design the table can generally be moved out of the working chamber and on to a setting up platform to enable it to be loaded, for example, by a crane, or to enable the workpieces to be changed more comfortably (Fig. 187). This possibility of not needing to mechanically or electrically decouple either the working table or any other type of positioning equipment is of particular importance in the case of program controlled welding in order to ensure reliable transfer of measurements of the workpieces' position.

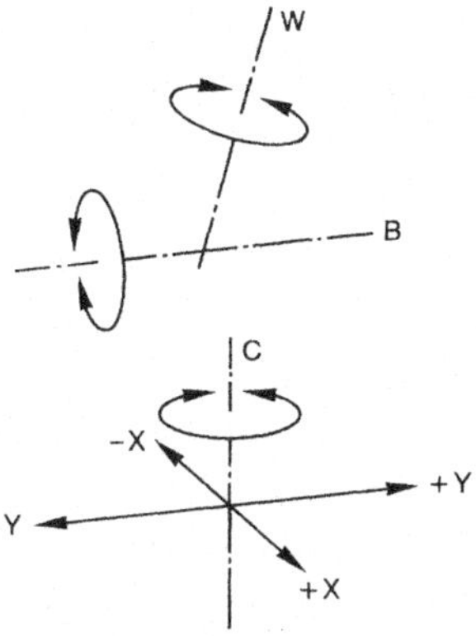

Fig. 201 The axes of movement of the arrangement shown in Fig. 200: B – Tilting axis; C – Swivel axis, W – Axis of rotation, X;Y – Axes of translation.

Fig. 200 Multi-axis rotary positioning system.

Fig. 202 Rotary positioning system with free end support, of 0.5 tonne capacity.

For welding circumferential seams there is equipment which can be mounted on the working table which has either vertical or horizontal rotational axes (Fig. 197, centre and right) or which can tilt the rotational axis into various positions (Fig. 200 and 201). The load carrying capability of the positioning equipment must of course be adapted to suit the weight of the workpiece to be welded (Fig. 202).

In standard equipment this may be from 0.15 up to 5 t for traversable tables and from 0.1 to 2 t in the case of rotational equipment with horizontal axes. Depending upon the actual welding operation, the displacement transducers on the working table and the rotational arrangement are able to measure the position of the workpiece to within the micrometre range. Here it is a definite advantage that, in contrast to machine tools, in electron beam welding positioning is not subjected to any load by the operation being carried out.

8.5 High voltage supply and control systems

A high voltage power supply is required to provide the accelerating and control voltages required by the electron beam gun, as well as to provide the heating current to the cathode. The high voltage supply consists of a transformer, the primary windings of which are connected to the 380 V three phase mains supply and the secondary windings of which supply the alternating voltage produced to a rectifier (Fig. 203). The rectified high voltage is still not smooth enough, however, for the electron beam welding process. Additional methods of smoothing are thus employed depending upon the beam power and quality requirements of the welding process. Finally, two additional voltage sources are added to the accelerating voltage to provide heating for the cathode and for the beam current (control voltage). A three core cable (Fig. 204) connects these three high voltages from the high voltage power supply to the electron gun (Fig. 205).

To provide easy access the electrical components of the accelerating voltage, beam current and heating current control systems are located on top of the high voltage power supply itself or in the control cabinet, i.e. on the primary side of the transformer, whilst all the high voltage transmitting components are housed in an oil filled tank. The tank is filled with transformer oil which has a high degree of electrical resistance. The high voltage supply together with all the control instrumentation are sealed to prevent any possible danger to the operating personnel and to ensure that no damage can occur to other components of the machine, provided that the supply is used correctly and in accordance with the relevant VDE specifications.

The other control instrumentation, and in particular that for the lens current, movement of the workpiece and the PLC or CNC systems are arranged in a special cabinet at the side of the working chamber (Fig. 206 and 207).

8.6 Vacuum systems

8.6.1 General

In welding technology there are only a few other processes (diffusion welding, vacuum brazing) in which the point of welding is protected from the detrimental effects of atmospheric gases and vapours by use of a vacuum. In electron beam welding the vacuum also meets a number of other requirements.

In the electron gun it is indispensable for preventing oxidation of the cathode, for enabling the electrons to be collimated into a narrow beam and for preventing electrical discharge. In the working chamber the presence of a vacuum has the definite advantage of further enabling the electron beam to be focused down to the smallest diameter without any scattering effects, even over large

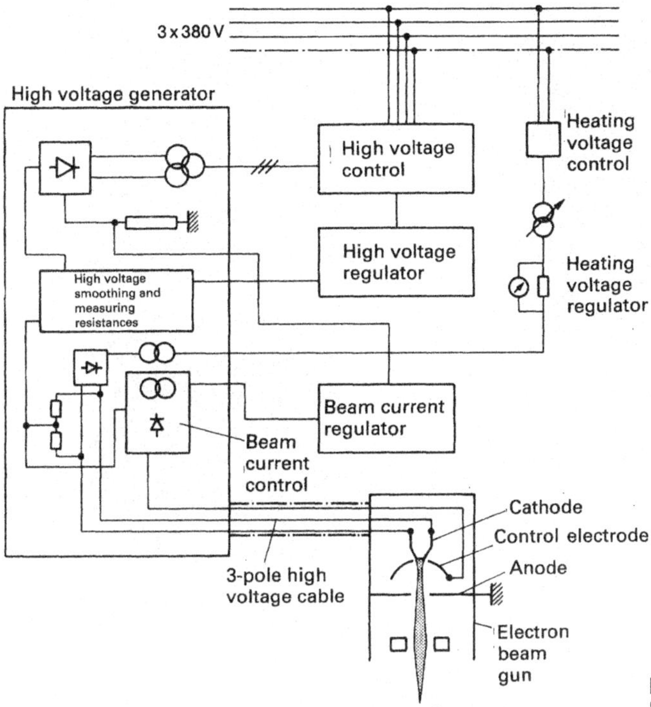

Fig. 203 Overview of a high voltage supply.

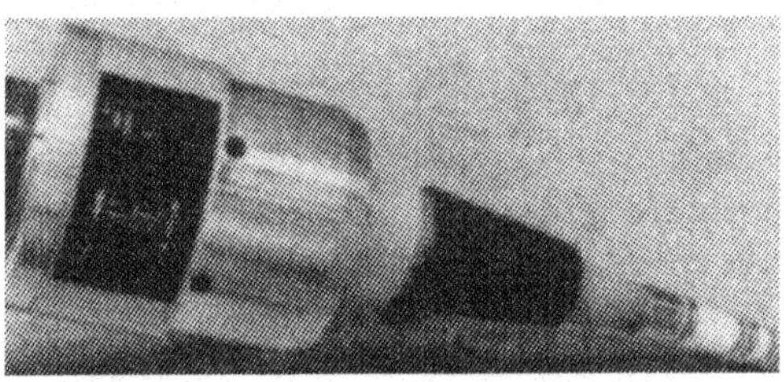

Fig. 204 The 3-pole connector of a high voltage cable.

Fig. 205 Electron beam gun showing connection of the high voltage cable.

Fig. 206 VDU and keyboard of a PLC system.

Fig. 207 VDU, keyboard and computer (from top to bottom) of a CNC system.

working distances. Vacuum systems are thus important component parts of all electron beam welding machines for a number of specific technical reasons.

8.6.2 Physical basics

The atmosphere which surrounds us is made up of gases (oxygen, nitrogen, etc) and vapours (water vapour). At normal ambient temperatures and pressures such vapours can be condensed, i.e. transformed into the liquid state, by a vacuum. The same cannot be achieved in the case of gases. Gases and vapours both consist of free atoms and molecules. Each of these particles of matter has a mass and moves at a speed which is temperature dependent. If they meet a barrier, e.g. the wall of a container, they impart energy to the surface by impacting against it and the sum of these impact forces over time and surface area is measured by the pressure of the respective gas or vapour. An important term in vacuum technology is the particle number density.

The standard pressure unit in vacuum technology according to DIN 28 400 is the millibar = mbar which is derived from the bar where 1 bar = $10^5\,N \cdot m^{-2}$. Normal atmospheric pressure is considered to be 1013 mbar. For operating electron beam welding machines soft vacuums are considered to be in the range p = 1 to 10^{-3} mbar, and high vacuums in the range p = 10^{-3} to 10^{-5} mbar.

Depending upon the pressure there are different physical laws governing the movement and mutual interaction between the gas molecules and atoms. For an understanding of the processes taking place in the vacuum in the working chamber it should be realised that the gas particles are not only hindered to a greater or lesser extent in their movement in space, but are also bound to the surface of solid bodies by adsorption effects. Desorbtion, that is 'sucking' during evacuation of these particles off the surface on to which they have been adsorbed is only marginally dependent upon the pressure, and to a much larger extent upon the time under vacuum and the surface finish and temperature of the wall of the chamber, the positioning equipment and the workpiece, etc. It must be realised that in the high vacuum range, i.e. $p < 10^{-3}$ mbar, more gas particles are to be found on surfaces than freely moving in space. Thus in electron beam welding machines a rough, dirty chamber wall, positioning equipment incorporating narrow gaps (e.g. motor windings) or a workpiece with a large surface area, can all considerably lengthen the evacuating time in spite of the use of powerful pumps.

8.6.3 Pumps

Evacuation of an enclosed volume involves forcing out air trapped within it against atmospheric pressure. For example, to achieve a vacuum of 10^{-5} mbar in the beam generator the air must be compressed beforehand by a factor of 10^8, which a vacuum pump alone is not able to achieve. Even backing of one pump with a second of the same type will not achieve success because with decreasing pressure the method of operation of the pumps used must be adapted to suit the physical laws and conditions predominating. For these reasons, the following types of pump are used in electron beam welding machines:

Mechanical pumps:	– Rotary vane pumps;
	– Roots pumps;
Kinetic pumps:	– Turbomolecular pumps;
	– Oil diffusion pumps;
Condensation pumps:	– Cryogenic pumps.

Figure 208 shows the respective working ranges of the individual types of vacuum pump.

In electron beam welding machines, only rotary pumps work directly against

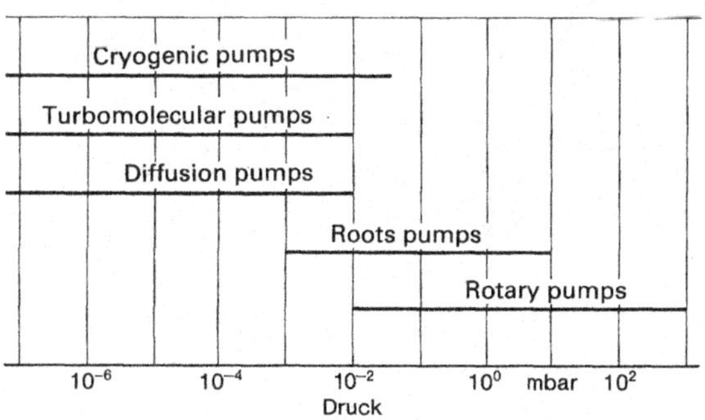

Fig. 208 The working ranges of the respective types of vacuum pump.

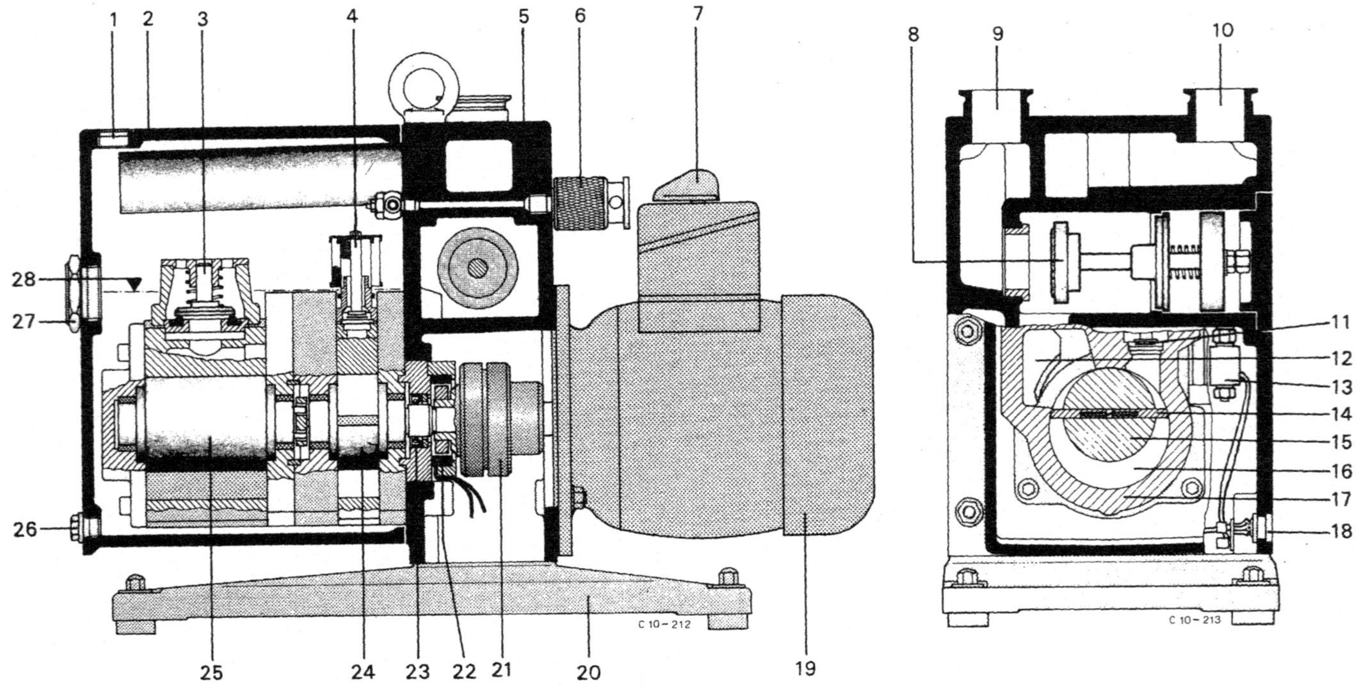

Fig. 209 Cross section of a two-stage rotary pump. 1 Oil filler plug; 2 Cap; 3 Over pressure valve; 4 Pump valve; 5 Supporting frame; 6 Gas ballast valve; 7 On and off switch with motor circuit breaker; 8 High vacuum safety valve; 9 Vacuum connection; 10 Exhaust vent; 11 Pump valve (as 4); 12 Suction channel; 13 Magnetic solenoid valve; 14 Rotor drive; 15 Rotor; 16 Working chamber; 17 Pumping cylinder; 18 Monitoring connection point; 19 Motor; 20 Baseplate; 21 Motor coupling; 22 Dynamo; 23 Radial shaft seal; 24 Stage 1 pump; 25 Stage 2 pump; 26 Oil drain plug; 27 Oil level glass; 28 Oil level.

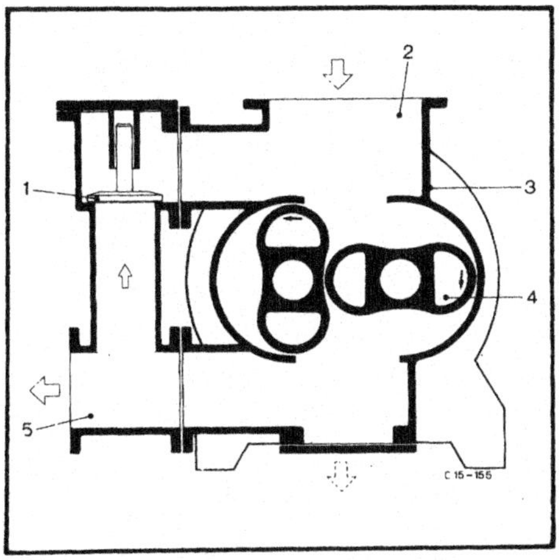

Fig. 210 Cross section of a Roots pump. 1 Over pressure valve; 2 Suction side; 3 Casing; 4 Rotary pistons; 5 Exhaust side.

Fig. 211 Roots pump of 1500 $m^3 \cdot h^{-1}$ nominal pumping capacity.

atmospheric pressure. They thus fulfill the function of rough pumping for each of the other types of pump.

Rotary pumps consist of two cylindrical casings each containing an eccentrically revolving rotor (Fig. 209). The rotor is fitted with a number of vanes pushed outwards by springs which wipe around the inside of the casing thus forming a sickle shaped evacuation chamber. As the rotor revolves, air is sucked into the chamber, forced towards the exhaust valve, compressed to atmospheric pressure and finally forced out into the atmosphere. Rotary pumps can achieve an end pressure of between 10^{-1} and 10^{-2} mbar.

Roots pumps, also known as Roots blowers, have two rotors housed in a figure of eight shaped chamber (Fig. 210). The rotors are interconnected by a toothed gear system in such a way that they are able to rotate past one another at high speed without touching. Roots pumps have a high suction capacity in spite of being dimensionally compact (Fig. 211) and as such are very suitable for rapid evacuation of large working chambers when welding is to be carried out at a pressure of about $p_A = 10^{-2}$ mbar. Depending upon the pressure at the exhaust side, a varying amount of air flows through the gaps back to the suction side. Rotary piston pumps thus achieve their maximum suction power when used in conjunction with a backing pump (rotary pump).

At pressures below 10^{-3} mbar (molecular flow), gas molecules and atoms are able mainly to move freely in open space. If under these conditions they collide with a solid which is also moving, then in addition to their thermal energy they will also be given an additional impulse (Fig. 212). If the particle subsequently collides with a second moving solid, the process is then repeated. Thus, from the random thermal movement of the gas particle prior to collision with the solid a directional movement is produced. The turbomolecular pump works exactly according to this principle. The pump consists of a rapidly spinning rotor and a stator both of which have numerous inclined vanes (Fig. 213 and 214). The particular advantages of turbomolecular pumps are, for mechanically operating pumps, their very low end pressure of about 10^{-10} mbar, their compact construction, their completely independent working position (either horizontally or vertically rotational axis) and the surety that there are no hydrocarbons (e.g. pumping oil) which could enter the vacuum. The suction power of such pumps is limited and thus they are mainly used together with a backing pump for evacuating beam generators.

The diffusion pumps in electron beam welding machines use a particular type of oil (with a low vapour pressure) as the transport medium.

The oil is heated to its boiling point, at which the oil vapour produced rises and flows at supersonic speeds in an umbrella shaped cloud out of a multi-staged, ring shaped system of jets and on to the cooled walls of the casing of the pump. Here it condenses and flow back to the heating vessel (Fig. 215). The gas particles on the suction side of the pump diffuse into the oil vapour umbrella and become attached to the heavy oil molecules. These then pass on downwards through the vapour umbrellas before condensing at the bottom of the column where the gas particles are released and extracted by the backing pump. A small amount of oil inevitably finds its way into the suction side of the pump although with use of refrigeration traps this can largely be suppressed. This must be taken into account when welding certain types of material (nickel alloys, special metal alloys etc). Diffusion pumps are robust and have a high suction capability at pressures of $p < 10^{-3}$ mbar. They are thus used extensively to evacuate working chambers in which welding is to be carried out at pressures of $p_A < 10^{-3}$ mbar (Fig. 216).

Cryogenic pumps operate on the principle that gases condense on highly cooled surfaces. The temperatures achieved in modern machines of from 10 to 20 K (−263 to −253 °C) cause water vapour and almost all atmospheric gases to condense and solidify; if required certain residual gases (helium and hydrogen) can be adsorbed using active carbon.

The refrigeration units incorporated in these types of machine need only be of relatively low power because of their highly effective insulation against

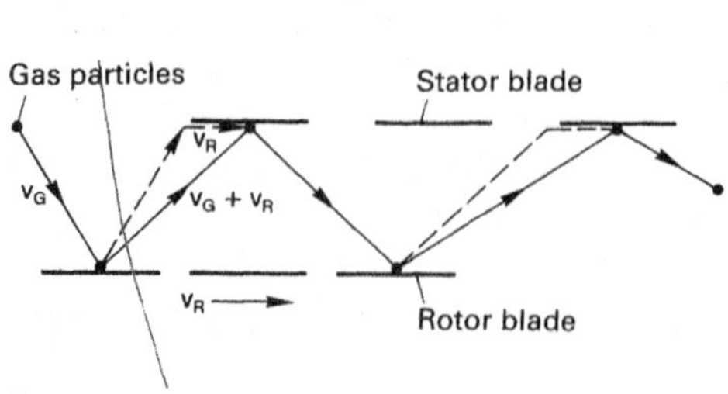

Fig. 212 Illustrating the gas particle acceleration occurring in a turbomolecular pump.

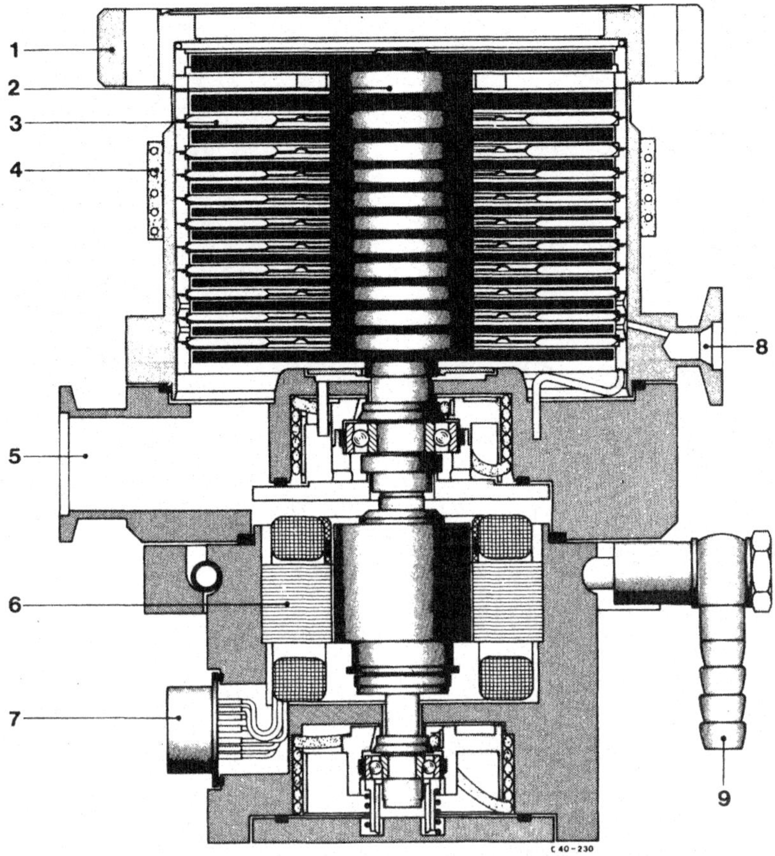

Fig. 213 Cross section through a turbomolecular pump. 1 UHV connection; 2 Rotor; 3 Stator; 4 Heating; 5 Roughing pump connection; 6 Motor; 7 Electrical connection; 8 Purging connection; 9 Cooling water connection.

Fig.214 The rotor blades of a turbomolecular pump.

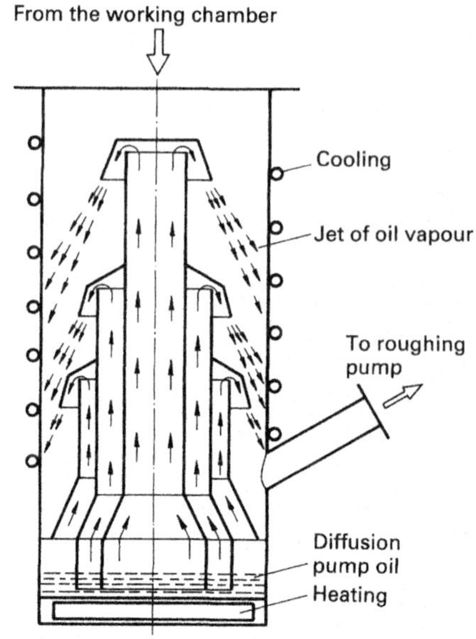

Fig. 215 The operating principle of a diffusion pump.

Fig. 216 Oil diffusion pump of 3100 l·s^{-1} pumping capacity.

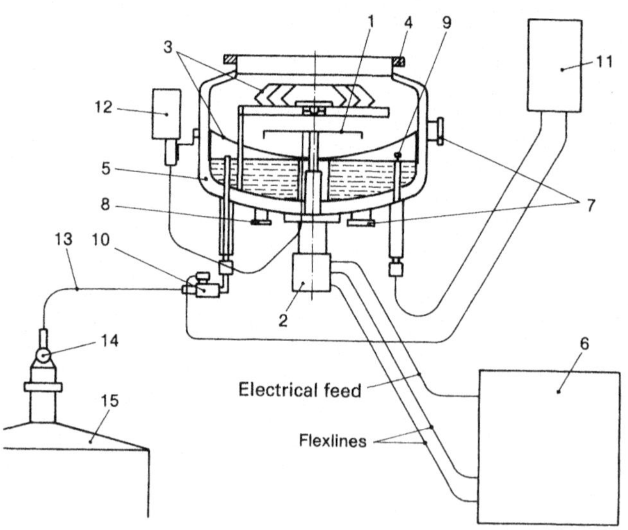

Fig. 217 A cryogenic pump using liquid nitrogen (LN_2) shielding. 1 20 K condenser with active carbon; 2 Cooling head; 3 77 K shielding with LN_2 container; 4 High vacuum connection; 5 Pump casing; 6 Compressor unit; 7 Roughing pump connection; 8 Connection point for measuring gauges and over pressure safety valve; 9 LN_2 level transducer; 10 LN_2 entry valve; 11 LN_2 control unit; 12 Temperature measuring unit; 13 LN_2 feed lines; 14 LN_2 lifting pump; 15 LN_2 storage dewar.

radiant heat. The insulation itself is cooled to about 60–100 K either by the first stage of the refrigeration unit (autonomous cryogenic pumps) or using liquid nitrogen (LN_2 = liquid nitrogen) stored in vacuum flasks (Fig. 217). Cryogenic pumps store the gases they extract, thus after a certain time (in the case of electron beam welding machines every 3 to 5 days) cooling must be interrupted (regeneration) to allow the frozen gases accumulated on the condenser surfaces to vaporise. Cryogenic pumps require a previously pumped rough vacuum of 10^{-2} mbar before they can operate, and can achieve end pressures of about 10^{-11} mbar. They not only ensure a vacuum free of hydrocarbons but also have a high suction capability in the pressure range $p = 10^{-2}-10^{-3}$ mbar (flow transition range) in particular for water vapour. They are thus significantly more efficient than either diffusion or turbomolecular pumps.

8.6.4 Instrumentation

Few methods of measurement can measure the pressure in a vacuum directly and independently of the type of residual gases remaining. Amongst these are vacuum gauges with membranes and piezo-resistive pressure gauges for use in the range 10^3-1 mbar.

All other methods of measurement, in the strictly physical sense, measure effects which are not actually pressure dependent, but are based upon the particle number density in the vacuum. These instruments must be calibrated for the particular type of gas to be measured (in the case of electron beam welding: air).

Pirani vacuum gauges measure the thermal conductivity of gases which decreases with decreasing particle number density (lower pressure). In the measuring tube there is a current carrying resistance wire made of a material with a high temperature coefficient. The respective thermal conductivity of the surrounding gases determines the temperature of the wire and thus its electrical resistance, which forms part of a Wheatstone bridge. This type of instrument is capable of measuring pressures in the range $1-10^{-3}$ mbar.

Methods used for measuring lower pressures are based upon ionisation of molecules and atoms. In instruments working on the Penning principle, a gas discharge is produced between a heated cathode and anode by applying a direct voltage of about 3 kV. The current which flows between the electrodes is a measure of the particle number density (and from this the pressure) of the surrounding gases. Cold cathode vacuum meters are very robust and not sensitive to air leakages. Their measuring range ends at about 10^{-10} mbar. The measuring range of heated cathode vacuum meters is down to 10^{-9} mbar.

8.6.5 Construction and principle of vacuum systems

Both for physical and economic reasons, the short evacuation times required for electron beam welding can only be achieved using a number of different types of pump designed specifically to work together. Thus, modern electron beam welding machines generally have rotary and Roots pumps for use in the range from atmospheric pressure down to about 10^{-2} mbar, together with oil diffusion, turbomolecular and cryogenic pumps for the final range down to about 10^{-5} mbar. Figure 218 shows a vacuum circuit diagram which details the different stages in evacuation of the working chamber and beam generator of an electron beam welding machine from the start of the evacuation process.

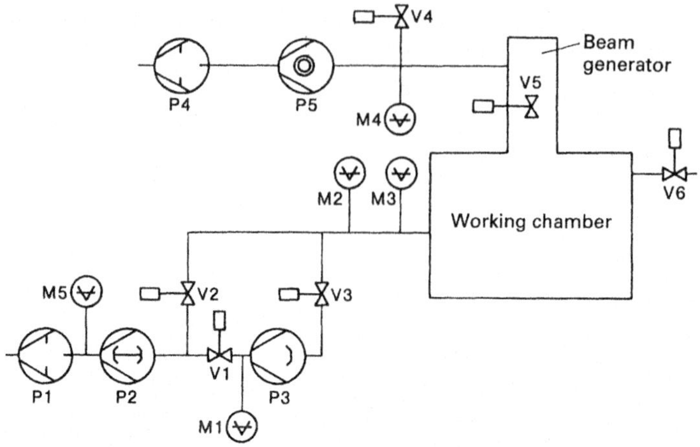

Work stage	Switching command
1	Evacuate the working chamber
1.1	Switch on pumps P1 and P2 Open valve V1 Close valves V2, V3, V5 and V6
1.2	Switch on pump P3 when measuring instrument M1 signals $p < 10^{-1}$ mbar
1.3	Close valve V1 and open valve V2 when measuring instrument M1 signals $p < 10^{-2}$ mbar
1.4	Open valves V1 and V3 and close valve V2 when measuring instrument M2 signals $p < 10^{-2}$ mbar
1.5	Working pressure is reached when measuring instrument M3 signals $P_A < 5 \cdot 10^{-4}$ mbar
2	Evacuate the beam generator
2.1	Switch on pumps P4 and P5, open valve V4
2.2	Working pressure is reached when measuring instrument M4 signals $P_E < 5 \cdot 10^{-4}$ mbar
3	Carry out welding
3.1	Open valve V5
4	Ventilate the working chamber
4.1	Close valves V3 and V5
4.2	Open valve V6

Fig. 218 Functioning of a standard vacuum system for electron beam welding machines. P1;P4 Rotary vacuum pumps; P2 Roots pump; P3 Oil diffusion pump; P5 Turbomolecular pump; M1;M2 Pirani gauges; M3;M4 Penning gauges; M5 Leakage location tube; V1 . . . V3 Valves; V4 Venting valve for the beam generator chamber; V5 Column value; V6 Venting valve for the working chamber.

If, as in the case of electron beam welding unalloyed or low alloy steels, a working pressure of $p_A = 10^{-1}–10^{-2}$ mbar is adequate, then the pump P3, valves V1 and V2 and measuring M1 and M3 will not be required. The column valve V5 is additionally equipped with a pressure stage and acts as an airlock between the working and beam chambers. The advantage of a having a higher

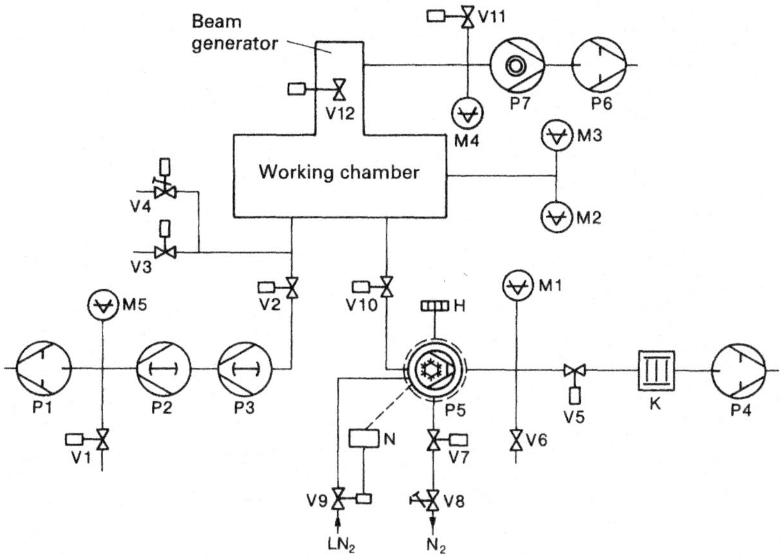

Fig. 219 Functioning of a vacuum system for electron beam welding machines incorporating a cryogenic pump. P1;P4;P6 Rotary vacuum pumps; P2;P3 Roots pump; P5 Cryogenic pump; P7 Turbomolecular pump; M1;M3;M4 Pirani gauges; M2 Penning gauge; M5 Leakage location tube; N Regulating system for the LN_2 level; V1 Emergency vent value; V2;V5;V7;V10 Valves; V3 Venting valve for the working chamber; V4;V8 Regulating valves; V6 Safety valve; V9 Regulating valve for the LN_2; V11 Venting valve for the beam generator; V12 Column valve; H Regenerator heater; K Catalytic trap.

pressure in the working chamber is that evacuation times are considerably reduced, this being especially useful in the case of batch type machines.

The provision of a separate vacuum pumping system for the beam generator is also useful when welding is to be carried out at pressures of from about $p_A = 10^{-1}$–10^{-2} mbar, allowing the beam generator to be evacuated without provision of a column valve V5. Two pumping systems operating independently offer the possibility of keeping the beam generator evacuated whilst changing the workpiece and of obviating the need to flood the chamber, which thus reduces the risk of contamination entering the system with the incoming air to a minimum (e.g. when changing the cathode). Also, by only having to evacuate the working chamber when changing the workpiece, a considerable decrease in pumping time is achieved.

Turbomolecular pumps are generally preferred for evacuating the beam generator. They can be used in all positions, such as when welding with different electron beam axes, and guarantee a dry, oil free vacuum. In contrast, for evacuating the working chamber, diffusion pumps are mainly used because of their much better suction power to price relationship. Only when a completely oil free vacuum must be guaranteed in the working chamber, e.g. for electron beam welding of certain high alloy steels, nickel alloys, special alloys, etc, must the diffusion pump be replaced by a cryogenic pump (Fig. 219). In

Fig. 220 Cryogenic pump mounted on the working chamber of an electron beam welding machine. Nominal pumping capacity: 19000 $l \cdot s^{-1}$ air or 48000 $l \cdot s^{-1}$ water vapour.

this vacuum circuit plan, a rotary pump P1 and two Roots pumps P2 and P3 have been included for rough evacuation of a relatively large working chamber ($V = 10\,m^3$).

At a pressure of 10^{-2}–1 mbar these are supplemented by switching in a further rotary pump P4, backing up a pre-evacuated cryogenic pump P5, through valve V10. The catalyser trap K prevents flow of any oil vapour back from the roughing pump. In contrast to autonomous cryogenic pumps, in which the insulation is cooled directly by the refrigeration unit, the cryogenic pump in Fig. 219 requires use of liquid nitrogen to increase the suction power. The sensor N checks the level of liquid nitrogen in the cryogenic pump and adjusts it by taking in or releasing liquid nitrogen through the valves V7 to V9. A heating strip wrapped around the pump, (Fig. 220) indicated by H in Fig. 219, is included to accelerate the regeneration cycle of the cryogenic pump. All the other components in this vacuum system operate in the same way as has already been described for Fig. 218.

8.6.6 Pumping times, leakage rates, locating leaks

The time taken to evacuate the working chamber is an important measure of the economic efficiency of an electron beam welding machine. It depends mainly on the size of the working chamber, the power of the pump and the working pressure to be achieved. For a clean, dry and empty working chamber, a defined pumping time is often quoted as part of the machine specification (CDE condition: clean, dry and empty) agreed between the user and the manufacturer.

During welding, the pumping time is subject to various influences amongst which are the size, cleanliness and condition (narrow gaps) of the surface of the workpiece and the clamping arrangement, and the time between venting and evacuating the working chamber, etc. Unduly long pumping times may also be caused by inadequate servicing of the vacuum pumps and leaks in the machine. Leaks, for example, occur in the seals on the doors to the chamber and may be a result of either general contamination or of work on the machine. To obtain a general view of the overall gas tightness of a welding machine it is recommended that the increase in pressure is measured during a long pause in operation. From the difference in pressure measured Δp_A, the volume of the working chamber V including the volume of the part of the gun up to the column valve (Fig. 188) and the time between the two pressure measurements, a leakage rate Q can be calculated. In perfectly operating electron beam welding machines this should be $< 5 \cdot 10^{-2} \text{mbar} \cdot \text{dm}^3 \cdot \text{s}^{-1}$. The leakage rate for the beam generator can also be measured in the same way.

If the leakage rate is significantly above this limit, the source of leakage should be found and sealed. This is also the case when welding gas sensitive materials such as titanium and zirconium where, because of the much more powerful vacuum pumps used, the presence of a leak may not show in longer pumping times and the weld pool could therefore come into contact with the incoming air.

In such a case the manufacturer of the machine should be called in with the various items of specialist equipment needed for locating leaks.

8.7 Examples of machines and welding jigs

Electron beam welding machines not only differ in their maximum beam power, but also in the dimensions of their working chambers and the way in which the workpiece is moved. For example, universal welding machines, as has already been mentioned, are equipped with working chambers of different standard sizes of 0.5–40 m^3 depending upon the purpose for which they are to be employed (Fig. 221 and 222). The load carrying capacity of the working table and, if used, the rotary positioning arrangement is designed according to the volume of the working chamber, which governs the maximum weight of workpiece which can be accepted (Fig. 223). Universal machines generally have an optical viewing system, manifold beam controls and powerful pumps to evacuate the larger working chambers rapidly. To add to the flexibility of these machines, the electron beam gun can generally be mounted on the working chamber in a number of positions (Fig. 72).

For welding small series of turned components in universal machines a multi-workpiece positioning arrangement is used, in order to weld the greatest possible number of workpieces in a single evacuating cycle (Fig. 224).

Large numbers of turned components, such as toothed gear wheels in the gearbox manufacturing sector of the automobile industry, are welded on cycle type machines. The particular characteristics of these types of machine are their small working chambers, which can accept only one or at most a few workpieces together with their clamping jigs at any one time. These machines have working chambers with volumes of 1–20 litres which makes it possible to

Fig. 221 Electron beam welding machine for welding various types of small component. P = 3 kW, V = 0.6 m^3.

Fig. 222 Production electron beam welding machine for manufacturing gas turbine rotors. P = 30 kW, V = 20 m^3.

Fig. 223 Electron beam welding machine for welding in the aerospace industry. P = 2 × 30 kW, V = 18 m^3.

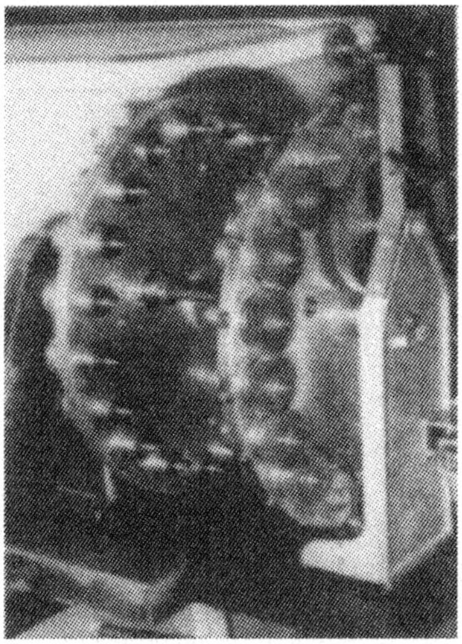

Fig. 224 Multi-component positioning system for welding small series of rotating components.

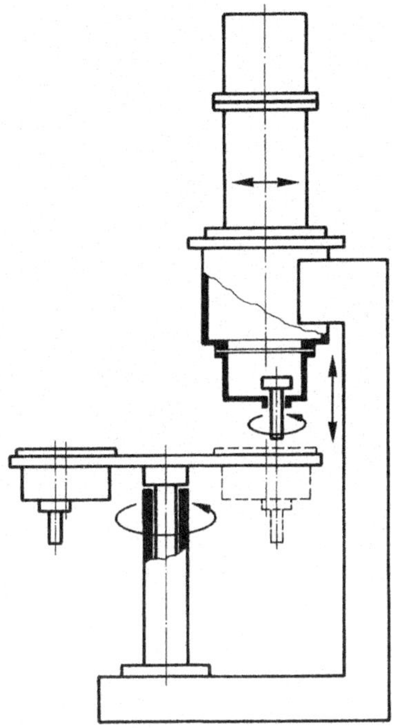

Fig. 225 A production line welding machine.

Fig. 226 A cycle type machine for welding gear components for the automotive industry.

Fig. 227 A production line machine for welding parts for the measuring and control industry.

Fig. 228 An arrangement for automatically feeding a production line welding machine for toothed gear wheels.

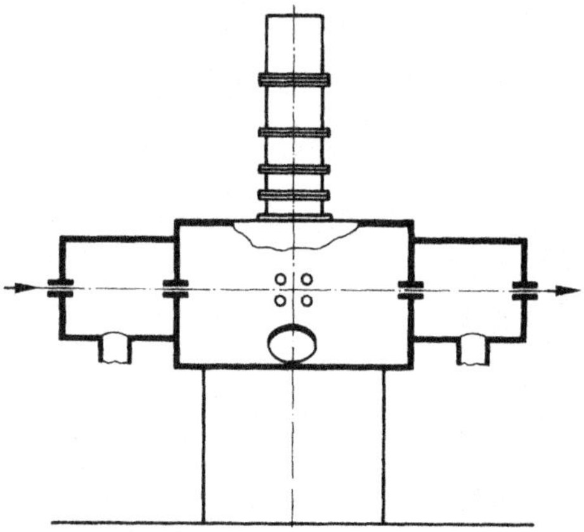

Fig. 229 Construction of a continuous machine for welding strip material.

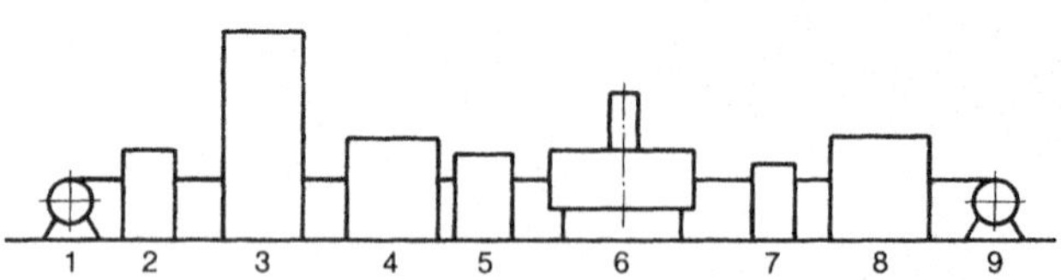

Fig. 230 A complete installation for continuous welding of bimetallic band saw blading material [142]. 1 Reel of unwelded strip; 2 Machine for flash butt welding the ends of the strip; 3 Strip storage hopper; 4 Degreasing plant; 5 Washing and drying plant; 6 Electron beam welding machine; 7 Strip counting unit; 8 Stress free annealing stage; 9 Take up reel for the welded bimetallic strip.

Fig. 231 Part of a machine for welding bimetallic band saw blading material.

Fig. 232 Electron beam welding machine with pressure stage airlocks for welding pre-combustion chambers for diesel motors.

evacuate them in only a few seconds (Fig. 225). In the case of cycle machines no optical viewing system is necessary and the operating console will only have switches to turn the machine on and off, an emergency off switch and a number of indicator lamps to monitor (Fig. 226 and 227). The entire working sequence, from evacuating, rotating the workpiece, welding and changing the workpiece is all carried out automatically according to a predetermined program and with a high degree of reproducibility of the set welding parameters. In incorporating such machines into transfer production lines, special additional pieces of equipment automatically transport the workpieces to the point of welding and move them on once this operation has been completed (Fig. 228).

Continuous welding machines, representing another type of electron beam welding machine, are mainly used to manufacture bimetallic semi-finished products such as bandsaw blades, thermo-bimetallic materials, electrical contacts, etc.

The workpiece is fed into the working chamber in the form of an endless band of material through a special arrangement of vacuum locks, where it is continuously welded and then transported through further locks out of the machine (Fig. 229). Depending upon the dimensions of the band, welding is carried out at speeds up to $330\,mm \cdot s^{-1} = 20\,m \cdot min^{-1}$. This puts a severe requirement on the dimensional tolerances of the workpiece as well as on the constancy of the welding parameters. For continuous welding of bimetallic strip a row of other work stations is connected to the continuous welding machine (Fig. 230). Amongst these are a butt welding machine to join the strips end to end to form an endless coil, together with various types of cleaning plant, numerous types of measuring and monitoring arrangements and, for example, for band saw blading, an induction furnace for stress relief annealing of the strip prior to mechanical working (Fig. 231). There is more on bimetallic band saw blading in section 10.4.

Other continuous welding machines operate with pressure stage airlock systems and are suitable for individual workpieces, mostly turned items with small dimensions (Fig. 232). The workpieces are fed into the working chamber through long airlock tubes and after welding are brought out of the chamber through a similar system into the surrounding atmosphere. A well known example of this is the electron beam welding of pre-ignition chambers for diesel motors (see section 10.2).

9 Quality levels and acceptable variations in electron beam welds

9.1 General

Before an electron beam welded component is put into service it is tested according to the nominated specification. Testing means measuring the item, comparing the values obtained with a given standard and evaluating the results. In this case the given standards represent the characteristics of a 'normal' electron beam weld. In evaluating the results it is determined whether the results of the measurements made lie within the acceptable variations or tolerances laid down in the standard. In other words, the description of the particular characteristics and acceptable variations is simply a quantitative statement of what defects in an electron beam weld may be regarded as acceptable.

In general in electron beam welding, as in all other welding processes, a verified set of results must be available if a guarantee is to be given as to the quality of any welding. In Germany this important matter is dealt with in the standards DIN 29 595 (aerospace industry) and DIN 8563 Part 11 (other industries), which specify the various quality classes, particular characteristics and acceptable variations for electron beam welding (see section 12).

DIN 8563 Part 11 applies to welds in steel in the thickness range 1–50 mm, whilst in contrast DIN 29 595 applies to steel, aluminium and titanium alloys of 0.4–20 mm thickness. Both standards are the result of many years of intensive exchange of experience between users, machine manufacturers, research institutes and technical inspection bodies.

The following explanations are based mainly upon the information given in DIN 8563 Part 11, which is the standard with the broadest field of application. It should again be mentioned that the schematic diagrams of welds used to help clarify the technical details discussed in the following sections are not true to scale.

9.2 Evaluation groups

DIN 8563 Part 11 specifies three types of weld which are given the following identification:

AS – EB
BS – EB
CS – EB

A, B and C differentiate between the different quality classes: A is the highest and C the lowest. S denotes the 'butt' weld seam geometry and EB – to differentiate it from other welding processes – electron beam welding. Fillet welds (K), as has already been mentioned in section 6.3.2 are a special case and are not dealt with in this standard.

The starting point for explaining the quality groups is the class BS – EB, which is referred to below simply as class B. In the quality group class B, as has already been mentioned, the characteristics of a normal electron beam weld are summarised, that is of a weld produced in accordance with normal welding practice. This of course does not mean a weld completely free of all defects. In other words, the standard does not attempt to define what should be understood by normal welding practice. This would, however, be a finished weld with a certain degree of excess weld cap profile, undercutting, etc, and including the acceptable variations from specified values.

According to the agreed understanding of the specialists involved in writing the standard, an electron beam weld produced to the tolerances and characteristics of a class B weld should be able to be produced using normal welding practice. If a standard of welding exceeding normal welding practice is employed, then the tolerances and requirements detailed in class A should be fulfilled, whilst for welding carried out to a standard considered not consistent with normal welding practice, the tolerances and requirements detailed in class C would be acceptable.

DIN 8563 Part 11 and DIN 29 595 are standards which are manufacturing orientated. In a way similar to the more familiar dimensional tolerances, they specify variations for certain relevant factors, depending upon the quality class required, with which the design and production engineer are able to work. The problem is to be able to construct the component according to the specified requirements whilst taking into account these actual manufacturing 'welding tolerances'.

The considerations which led to the writing of DIN 8563 Part 11 and DIN 29 595 were purposely explained in some detail. The difficulty in setting up such a catalogue of defects is to determine the criteria for defining the term 'acceptable defect'. What is acceptable, and for what and to whom is it acceptable? If acceptable defects are specified without these being relative to variations from a given standard, then without further explanation such an evaluation would often be somewhat arbitrary.

9.3 Post-weld machining

If particularly high strength requirements are placed on a given electron beam welded component, then under certain conditions, even in class A, the variations occurring may be unacceptable. For example, for dynamic applications the welds should be free of all undercut, spatter and droplet formation along the root of the weld.

In these cases the possibility of carrying out post-weld machining should be mentioned, although this must be taken into account before welding when choosing the shape of the weld and the workpiece preparation (see Fig. 115 and 116). In the production documentation it must be remembered that any post-weld machining required for characteristics and variations not otherwise contained in quality class A must be detailed and specified separately.

9.4 Weld width

In contrast to standards and specifications for other welding processes, in DIN 8563 Part 11 and DIN 29 595 no maximum permissible width of the weld is specified. It has been shown that as far as assuring the quality of the weld is concerned, there is no reason to specify tolerances for this parameter. In practice, in most cases electron beam welds are not wide enough to have an adverse effect on the quality characteristics specified. A minimum width of 2 mm is, however, recommended for practical manufacturing reasons, such as the tolerances applied in the machining of the components, the accuracy of guidance of the weld positioning arrangement, and the accuracy of positioning of the electron beam, etc.

9.5 Excessive weld profile (convexity)

When in DIN 8563 Part 11, for a weld thickness of 10 mm, a maximum weld profile (convexity) for the crown and root of the weld of $\Delta a_1 = 2\,\text{mm}$ is specified, the question often arises as to why this profile should not be permitted to be even greater, and the disadvantages were this value to be exceeded. Figure 233 illustrates why the profile Δa_1 should not be allowed to be greater when the width of the weld b remains constant.

The reason for this limit is not the volume of the crown or root of the weld, but the angle α between the surface of the workpiece and the tangent to the profile of the weld. It is apparent that too small an angle α, even when $> 90°$, together with the undercutting which also often occurs, will cause a reduction in the strength of the weld. Since the width of the weld varies only slightly with the thickness, it is more practical to limit the profile of the weld. The maximum acceptable value Δa_1 is independent of the thickness of the weld and in all quality classes is not permitted to exceed 5 mm. In general, in actual manufacturing practice no real difficulty is experienced in meeting this requirement.

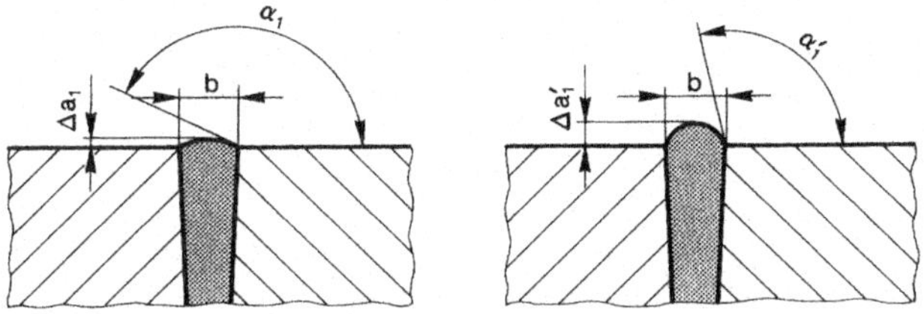

Fig. 233 The effect of weld convexity Δa_1 on the angle α between the surface of the workpiece and the tangent to the weld for a given weld width b.

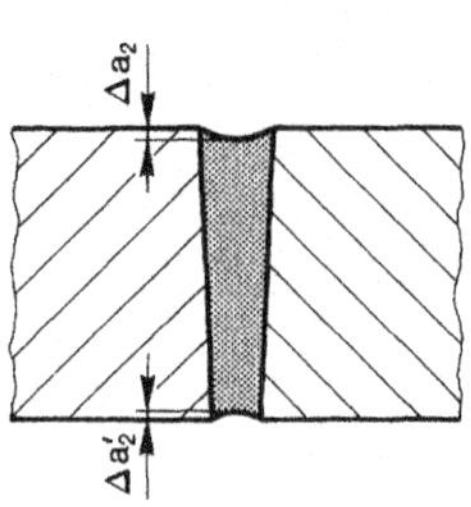

Fig. 234 Weld showing concavity Δa_2.

9.6 Weld concavity

With careful mating of the faces to be joined and choice of welding parameters, the transverse shrinkage which occurs across the weld can produce significant weld convexity.

If the weld gap b_s is unacceptably large, however, see Fig. 234, or for example welding is carried out with unsuitable weld parameters, concavity in both the crown and root of the weld or excessive droplet formation along the root can result. These types of defect can be extremely detrimental to the strength of the weld. It is thus understandable that such defects are judged very critically and are not permitted under any circumstances in class A welds.

If weld concavity cannot be avoided, then post-weld machining will be required which will need to be taken into account when choosing the shape of the weld seam by the provision of a suitable machining allowance (see section 6.3.1). In class B welds, however, the 'normal' weld is permitted to exhibit weld concavity Δa_2 up to 0.5 mm in welds less than $t \geqslant 15$ mm in thickness, and in thinner welds correspondingly less.

9.7 Undercutting

Undercutting can occur in all welding processes. The cause of undercutting is a complex interaction of, on the one hand, the surface tensions in the weld pool and, on the other hand, the movements of the weld pool and the solidification processes occurring. Whilst undercutting cannot always be prevented in electron beam welds, it must be judged particularly critically because of the often particularly sharp notch formed (Fig. 235). In class A and B welds DIN 8563 Part 11 permits a maximum depth of undercut of 0.5 mm for weld thicknesses of $t \geqslant 10$ mm. If, for example, a class A weld has particularly high strength

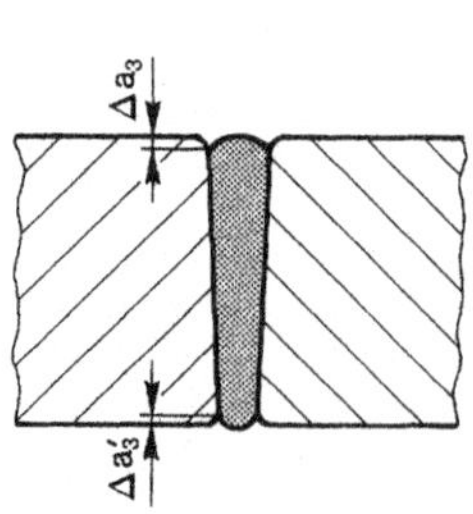

Fig. 235 Weld showing undercutting Δa_3.

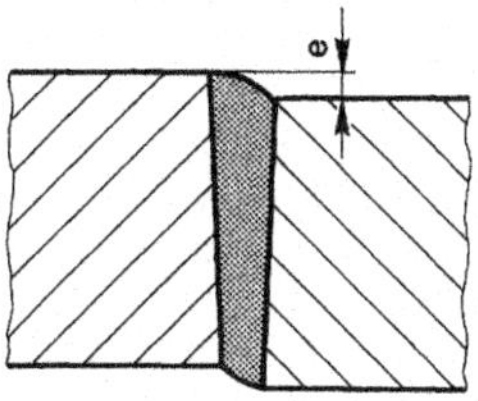

Fig. 236 Weld showing edge misalignment e.

requirements placed on it, then the weld settings governing beam oscillation will need to be optimised in a series of trial welds. On the other hand, with choice of a suitable weld shape and machining allowance, post-weld machining may be carried out instead.

9.8 Edge misalignment

The cause of edge misalignment lies not in the welding process but in the preparation of the workpiece for welding and in the construction of the welding jig. These, therefore, are the factors which must be considered in order to prevent or reduce the possibility of edge misalignment. The capability of the electron beam welding process to bridge, i.e. weld across, excessive misalignment is very limited (Fig. 236). In class B welds, depending upon the thickness of the weld, edge misalignment up to a maximum of 2 mm is accepted (e.g. for t = 10 mm a misalignment of e = 1 mm is allowed). For this reason, the numerous types of weld seam incorporating self-centering are used in manufacturing wherever possible (see section 6.3), to ensure that, taking into account the dimensional tolerances of the workpiece and its functions, no edge misalignment can occur.

9.9 Pores, shrinkage cavities

As explained in detail in sections 3.3 and 5.2.3, gas reactions involving accompanying or alloying elements, together with irregularities occurring during solidification of the mobile weld pool are the main cause of pores and shrinkage cavities in the weld. A large number of effective methods are available to the practical welder to prevent, or at least almost completely suppress, formation of such voids. Unacceptably high costs would be incurred, however, to produce a weld specified as being absolutely free from all pores and shrinkage cavities. It is much more economical, even in class A welds, to accept a limited number of such voids. If necessary, the effect which such pores and shrinkage cavities have on the properties of the manufactured product can be investigated experimentally [143] using artificially produced defects. In these investigations it has been shown that at least two methods of non-destructive testing, such as radiographic and ultrasonic testing, should be employed adequately to determine the size of a void.

To ensure the quality of a weld seam, standards and codes must set limits on the number and size of pores and shrinkage cavities it contains. Thus in

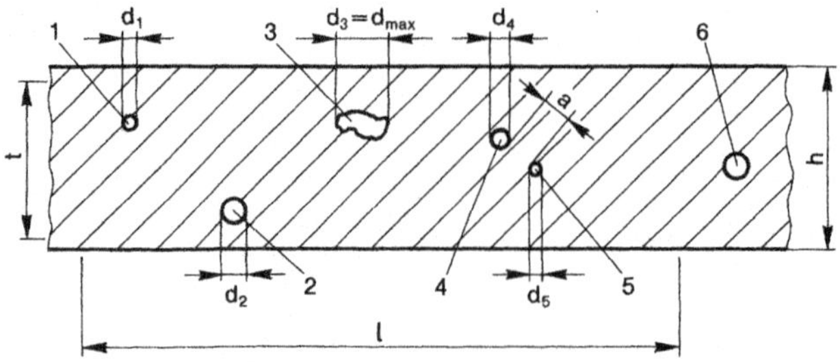

Fig. 237 The various dimensions of pores and voids found in welds. d_{max} the maximum longitudinal length of a void within the length of the weld l. a the minimum distance between two voids within the length of the weld l.

electron beam welding, in contrast to other methods of fusion welding, these types of defect are described in particular detail.

Within a certain length of weld one differentiates between the size of the maximum void, the sum of the surface area of all of the voids and the minimum distance between the individual voids, to be able to categorise both lines and clusters of pores (Fig. 237).

In class B welds, depending upon the thickness of the weld, an individual pore may have a diameter, or a cavity a length, of up to 3 mm. For the example of a weld of thickness t = 10 mm this limit would be 1 mm. It is further specified that in a length of weld l = 150 mm the sum of the surface area of all the voids present $f = \pi/4(d_1^2 + d_2^2 + d_3^2 \ldots)$ should not exceed more than 1% of the surface area of the longitudinal section through the weld times the thickness of the weld = $l \cdot t$. Figure 238 gives a visual impression of this numerical value. One would probably not accept this frequency of porosity in a radiographic film, which because it is irradiated in the thickness direction of the weld would not yield any information regarding the distribution of the porosity through the thickness. Nevertheless the size, frequency of and distance between these voids would be acceptable in a class B weld. With smaller weld thicknesses, however, the distances between voids are so small, even when in compliance with the maximum acceptable surface area of voids, that evaluation is carried out in terms of lines of pores which, having considerably more stringent requirements, would result in the weld being judged unacceptable.

From this it will be clear that the standards deal with the actual effect which individual voids resulting from the welding process have on the strength of the component. In the standards lines of pores, which are generally a result of poor preparation of the workpiece and in particular of the presence of impurities, are dealt with more stringently because of their crack-like nature.

9.10 Other points of note

Clearly fusion defects and cracks must be judged unacceptable in all classes of weld. Welds which have not been subjected to their specific in-service stresses must still be evaluated, and those containing these type of defects must be rejected or remedially repaired. Only in very special cases, with an exact knowledge of the behaviour of the material, the size of the crack or fusion

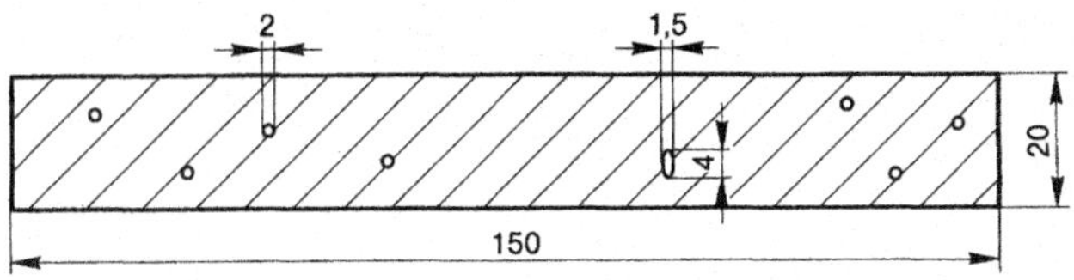

Fig. 238 An example of acceptable pore sizes and numbers according to the requirements of DIN 8563 Part 11 – BS–EB. The sum of the surface areas of the voids measures 1% of the longitudinal cross sectional area of the section of the weld.

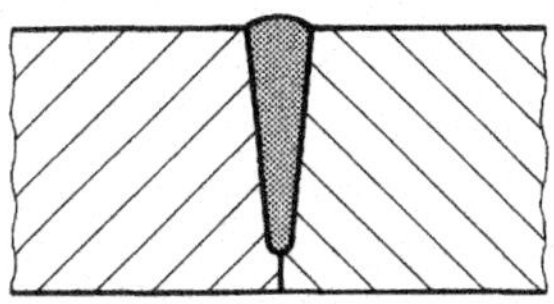

Fig. 239 The unwelded root of an I butt weld.

defect and the type of loading to be experienced in service, can any attempt be made, with the aid of fracture mechanics, to tolerate the presence of such defects [144]. This type of consideration is outside the scope of this book.

Lack of root penetration or fusion are also judged to be unacceptable in class A and B welds (Fig. 239). Exceptions are butt welds with centering lips (Fig. 111) in which the depth of the weld need not be the same as the full thickness of the workpiece.

Fused and adhering weld spatter is acceptable as long as it is of the same material as the parent metal. If there are particular requirements for the workpiece to be free of spatter, a note should be appended to the specified quality class to the effect that post-weld finishing is required. The same is true for metal sublimate. Metal vapour is formed, as has already been mentioned in section 6.13, even when the residual beam impinges upon the opposite surface of the workpiece or on the surface of the welding jig (e.g. in pipe welding), and it condenses on the surfaces of both the workpiece and welding jig. Metal vapour sublimate need only be removed in certain particular instances.

10 Examples of electron beam welded components

10.1 Jet engines, gas turbines

A need to weld titanium alloys, high temperature steels and nickel alloys, as well as requirements to save weight, to maintain close tolerances and to ensure high quality welds were the main reasons why, as long as ten years ago, electron beam welding was introduced into manufacture and repair of jet engines. The great majority of the work carried out is welding of circumferential seams, although radial seams are definitely preferred because of the unhindered shrinkage this allows which helps to suppress the tendency to cracking exhibited by many high temperature materials. Figure 240 shows a longitudinal section through a typical example of a drive shaft including four different welds. The shaft transfers up to 310 kW of power at a rotational speed of 12 000 rpm. The gear components and shaft sections are all made from the case hardening steel 15 NiCrMo 16, whilst the bearing journal is manufactured from the bearing steel X20 WCr 10 3. Figure 240 also contains details of the dimensional accuracies achieved. After electron beam welding the welds are machined and the gear components and bearing surfaces ground [145].

In addition to this particular example, Fig. 241 shows a welded pump shaft which has been sectioned to show details of the welding. Here as well use

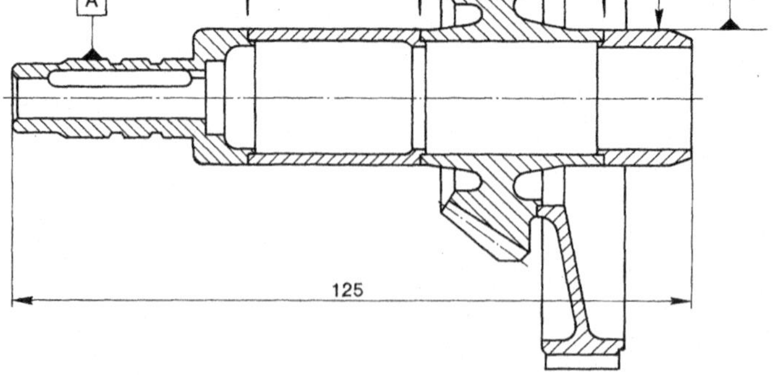

Fig. 240 A toothed gear wheel shaft with four butt joints prepared for electron beam welding. The numbers in boxes give the dimensional accuracy achieved after welding in mm [145].

Fig. 241 An electron beam welded pump shaft made of low alloy steel. External diameter ca 100 mm.

Fig. 242 An electron beam welded stator ring made of a titanium alloy. External diameter ca 600 mm.

of electron beam welding is particularly practical, because relatively costly individually machined parts can be welded together to form a complex finished component of high overall dimensional accuracy. Without this possibility, such a component would either need to be machined expensively in one piece from the solid, or the individual parts would be joined by screw threads.

Today, electron beam welding holds an outstanding position in the manufacture of stator rings (Fig. 242). The technique is used for joining the foot and shroud sections of the blades to one another and to the ring housing (Fig. 243). Here, the distortion free welding of a large number of seams arranged adjacent to one another is of particular importance. The high dimensional accuracy required can only be achieved with the use of CNC systems which, together with weld seam sensing systems employed, make electron beam welding of stator rings a completely automatic process operating to a particular preprogrammed welding sequence.

Welding of jet engine rotors requires both a high degree of dimensional accuracy as well as excellent weld quality and high reliability. Figure 244 shows

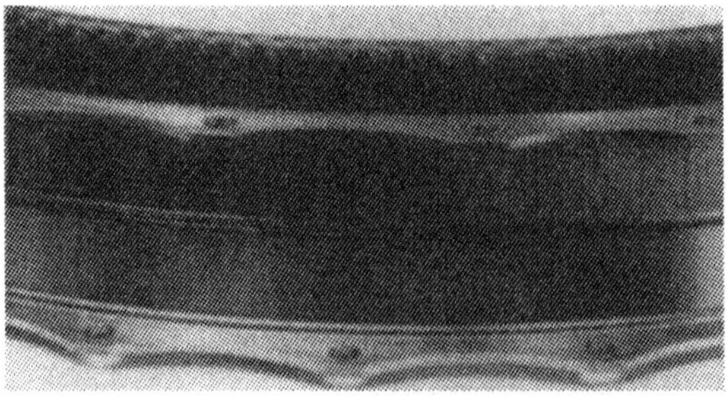

Fig. 243 Enlargement of Fig. 242, showing the stator blades welded together at the ring mantle.

Fig. 244 A segment of a rotor with two radial weld seams which have been post-weld machined. Material No.: 2.4668. External diameter: ca 400 mm.

a segment of a rotor made of a nickel alloy and incorporating two electron beam welds.

Here as well the component comprises a number of segments which are manufactured individually, thus providing savings in both labour and materials, prior to electron beam welding together. Because of the high rotational speeds and dynamic loadings to which this rotor is subjected, all the welds are machine finished.

The compressor rotor shown in Fig. 245 is about 1400 mm long and has a diameter of about 600 mm. It comprises nine individual discs welded together, four of which are made of a quenching and tempering steel (left) and five of a nickel alloy (right). Only electron beam welding can be used to weld the otherwise inaccessible circumferential seams (Fig. 246) whilst maintaining the close dimensional tolerances specified, as well as the dissimilar metal combination of a ferritic to an austenitic steel. The working chamber is arranged in such a way that all the circumferential seams can be welded in a single evacuation cycle. Figure 247 shows a similar rotor during its final ultrasonic inspection.

Fig. 245 A rotor made up of compressor discs of different materials ready for electron beam welding.

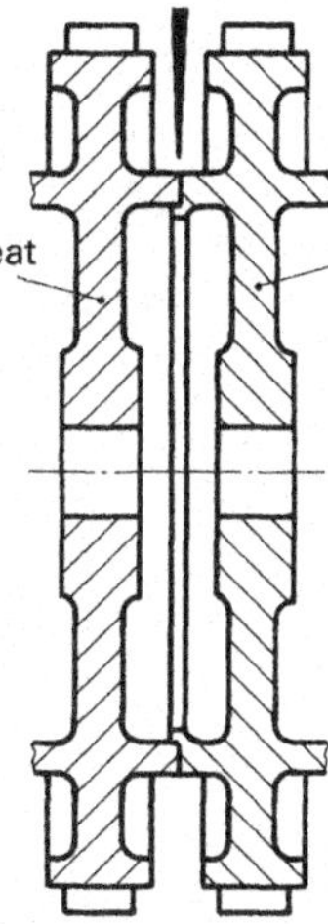

Fig. 246 Cross section through the compressor rotor shown in Fig. 245.

Fig. 247 Ultrasonically testing an electron beam welded rotor with automatic documentation of the test results.

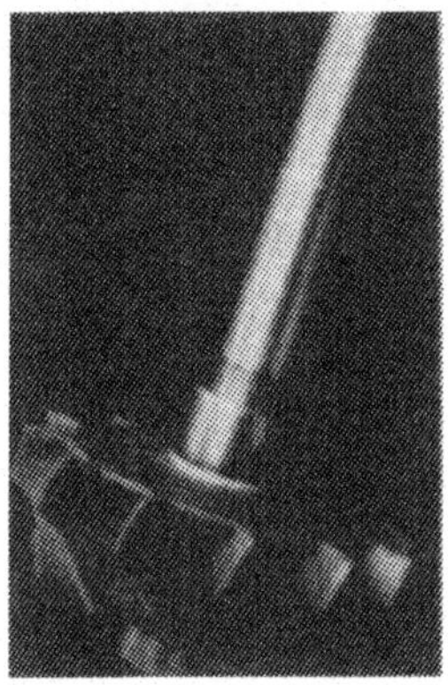

Fig. 248 A turbocharger made of a nickel based alloy and a free machining steel (shaft).

Another example of a similar combination of dissimilar materials being joined by electron beam welding is the turbocharger shown in Fig. 248. The impeller wheel is made of a high temperature nickel alloy and the shaft of a free machining steel, a combination of materials generally considered unsuitable for welding. With specific beam parameters, in particular a defined offsetting of the beam, it is possible to weld the parts achieving the degree of weld quality required consistent with the high mechanical and thermal stresses which this component experiences in service.

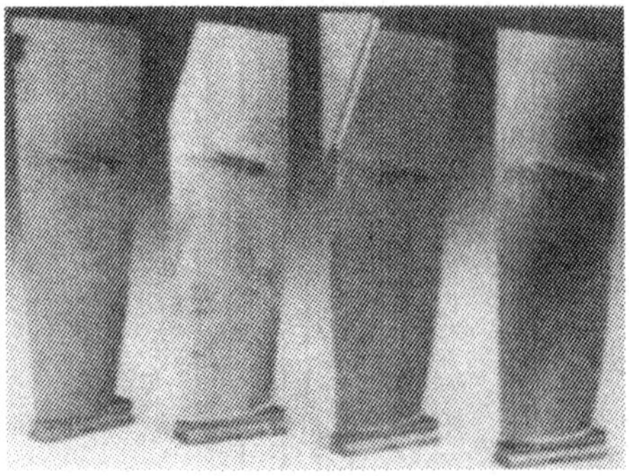

Fig. 249 The stages of remedial repair of compressor blades by electron beam welding: (from left to right) Blade damaged by foreign body impact; blade prepared for welding; electron beam welded blade; finally machined repaired blade.

Fig. 250 The high pressure casing of an aircraft jet engine.

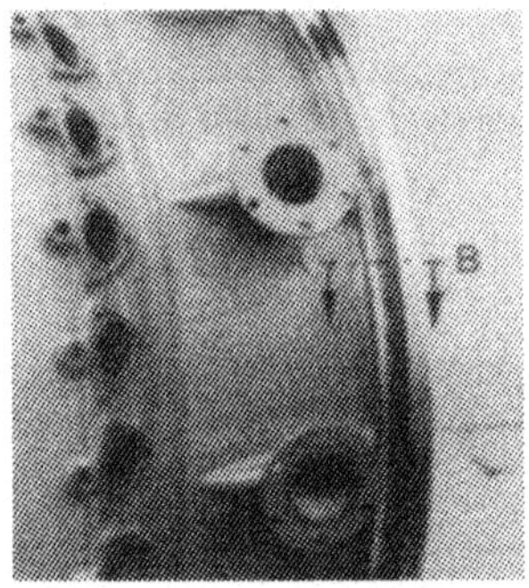

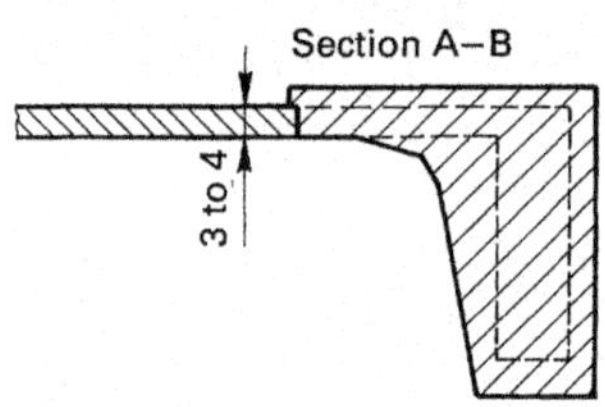

Fig. 251 Top: The flange of the casing shown in Fig. 250, repaired by electron beam welding. Bottom: The extra machining allowance on the welded flange.

Electron beam welding is also important in repair of jet engine components. In this respect the most frequent work carried out is repair of compressor fan blades which become damaged by foreign bodies. The blades are made of the forged titanium alloy TiAl 6 V 4 and because of their shape and size are extremely expensive. Figure 249 shows a series of such compressor blades in the damaged, cropped, as-welded and finished conditions. After heat and surface treatment a very acceptable weld quality is achieved which corresponds to about 95% of that of the parent metal. In addition to a saving in costs of several thousand DMarks per blade, repair of such blades by electron beam welding rather than purchase of new blades from outside companies also results in a considerable saving in time [146].

Fig. 252 Gearwheels prepared for electron beam welding. Left: Synchronous gear ring (top) and gear wheel (bottom) prior to assembly.

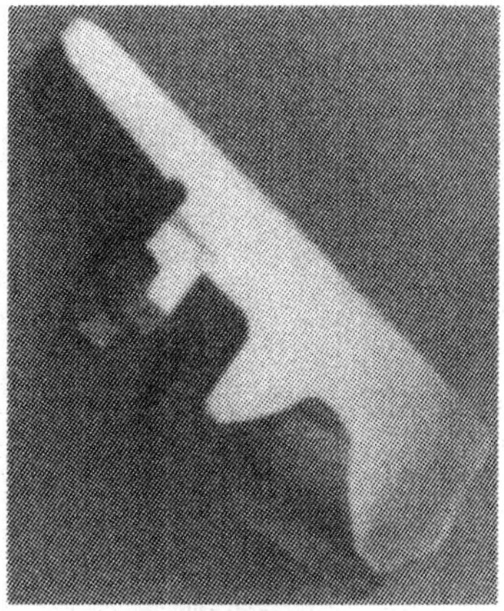

Fig. 253 Macrosection through an electron beam welded gearwheel with synchronous gear ring (1:1; Reproduced 0.45×).

After some 20 000 hours of service, fatigue cracks appear in the flange of a high pressure engine casing made from the nickel alloy 2.4668. The casing, which has a diameter of 1100 mm and wall thickness of 3–4 mm, is designed to operate at a maximum pressure of 30 bar (Fig. 250 and 251). The damaged flange is removed and a semi-finished profile welded on to the casing (Fig. 251, bottom). After machining, the jet engine casing is once again ready to be put back into service [146].

Other examples are welding of knife edges on to bearing seals and repair of magnesium alloy gearbox casings. Bushings and other inserts are also welded using a circumferentially deflected electron beam, without the need to rotate the workpiece itself.

10.2 Automobile industry

Another large field of use of electron beam welding in industry is welding of toothed gears for automobile gearboxes. As can be seen from the example shown in Fig. 252, electron beam welding is used mainly to join the synchromesh ring to the gear body. By being able to machine each of these components separately, significant savings in labour, materials and also in particular in the weight of the finished component can be achieved, see Fig. 156. If the welds are suitably positioned in the design of the component (Fig. 253 and 254) then no post-weld machining is necessary.

Special cycle type machines are now able to weld such gearbox components in a matter of seconds, and even interrupted welds can be precisely controlled according to the position around the circumference of the component (Fig. 255). Electron beam welding is also widely used for welding other gearbox

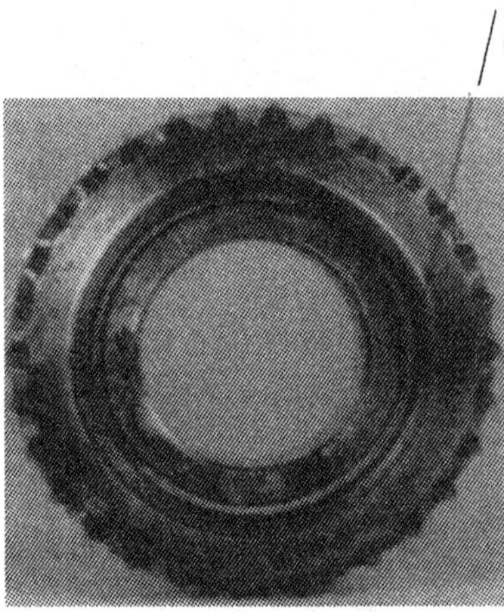

Fig. 254 An electron beam welded gearwheel (1:1; Reproduced 0.45×).

Fig. 255 Interrupted axial weld in an electron beam welded gearwheel (2:1; Reproduced 0.45×).

components. Figure 256 shows the drive-shaft of an automatic gearbox. The joint between the 20 MnCr 4 steel shaft and the RR St 4 steel planetary gear housing (Fig. 257) is electron beam welded: the thickness of the weld is 4 mm and the welding speed about $35\,mm\,s^{-1}$.

Trapezoidal lugs on planetary gear housings for commercial vehicles are also electron beam welded (Fig. 258, bottom, cross hatched). Because of the number required a universal welding machine is used, equipped with a rotating drum arrangement (Fig. 259). A drive motor rotates the workpiece into the welding position and also revolves it during the actual welding process. In this example the planetary gear housing is a forging requiring high dimensional accuracy across the width of the lugs as well as between the individual lugs themselves. By mechanically sensing the lugs, a very simple and effective method has been found of correctly controlling the position of the electron beam with respect to the workpiece in spite of the demanding tolerances required [147].

Fig. 256 Driveshaft with electron beam welded planetary gear cage.

Fig. 257 Macrosection through the weld shown in Fig. 256. Left: Planetary gear cage – material RRSt 4; Right: Driveshaft – material 20 MnCr 4. (40:1; Reproduced 0.3×).

Fig. 258 A planetary gear cage, as used in a number of commercial vehicles, prepared for electron beam welding. In each case one of the trapezoidally shaped lugs which are welded together is shown cross hatched. External diameters ca 120 and 150 mm [140].

The gearbox drive-shaft shown in Fig. 260 is also an interesting example of the possibility of saving both material and manufacturing costs by using electron beam welding. Without the axial circumferential seam shown in Fig. 261 the shaft would have either to be machined in one piece from the solid or the bell keyed to the shaft (probably necessitating lengthening of the shaft), and other additional steps would probably also have to be taken to transfer any possible axial forces. The photographs also impressively illustrate the highly finished condition in which gearbox components can be electron beam welded.

Electron beam welding is also used in the automotive industry for welding

Fig. 259 Drum rotating mechanism for electron beam welding of a number of planetary gear cages in one evacuation cycle [147].

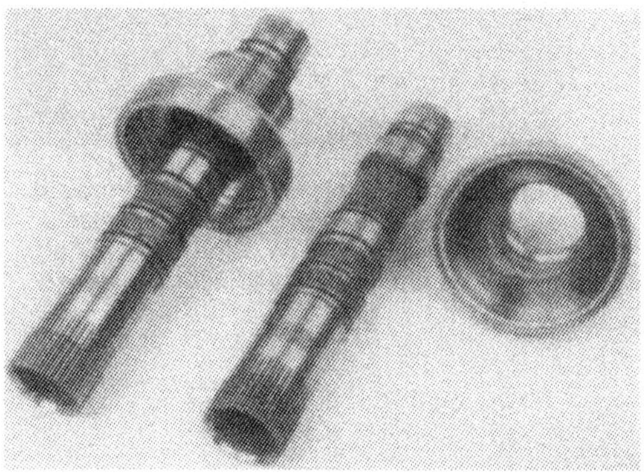

Fig. 260 A finally machined driveshaft prepared for electron beam welding (shaft ca 200 mm long).

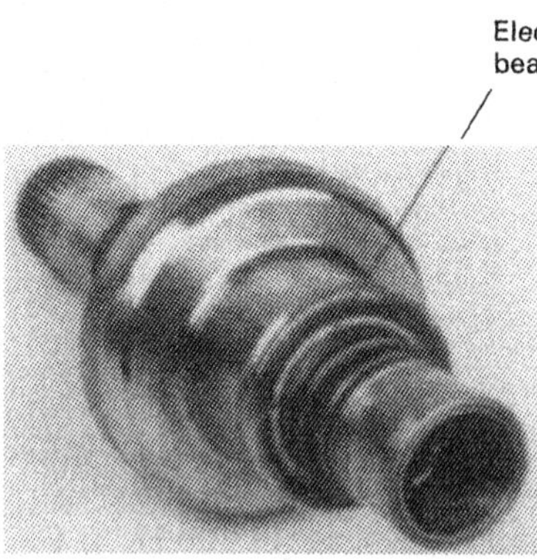

Fig. 261 The driveshaft in Fig. 260 showing the final axial circumferential weld.

axles. The special machine shown in Fig. 262 uses two electron beam guns to weld the brake flanges and axle tunnel on to each end of the axle bridge, this being accomplished in less than 7 min (Fig. 263). Of particular note is the brake flange weld seam. As has already been shown in Fig. 139, this incorporates a

Fig. 262 Machine (in background) with two electron beam guns, each of $P_{max} = 8.5\,kW$, used for welding rear axles for commercial vehicles. The assembly station is also shown in the foreground.

Fig. 263 An electron beam fillet weld on a brake flange (left) and radial weld in the axle tube (right).

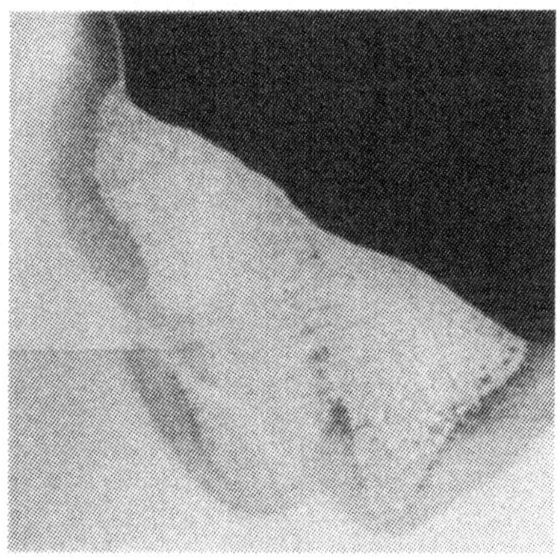

Fig. 264 Macrosection through the circumferential fillet weld shown in Fig. 263 (10:1; Reproduced 0.45×) [148].

step both to achieve the cross section needed to ensure the required strength (Fig. 264) and, together with the angled crown of the weld, to produce as uniform a stress flow as possible across the joint. Welding of this fillet weld with a 'real' a-dimension is possible only using a relatively large angle of beam deflection. To achieve this the two electron beam guns are arranged at an angle to the axis of the workpiece.

A similar machine is also used to weld the circumferential seams of the front axles for cars. In this case two cheeks forged from a quenching and tempering steel are joined to a central tube (Fig. 265). For this dynamically stressed component, the electron beam welds must have a relatively smooth surface

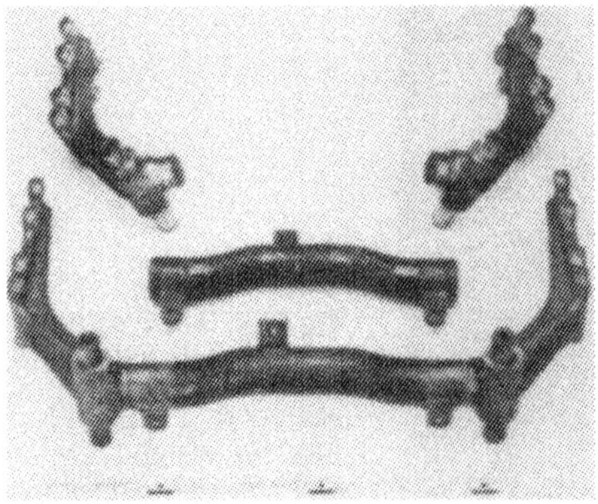

Fig. 265 Cross brace and front wheel suspension arms for a car manufactured with two radial circumferential welds (bottom) [148].

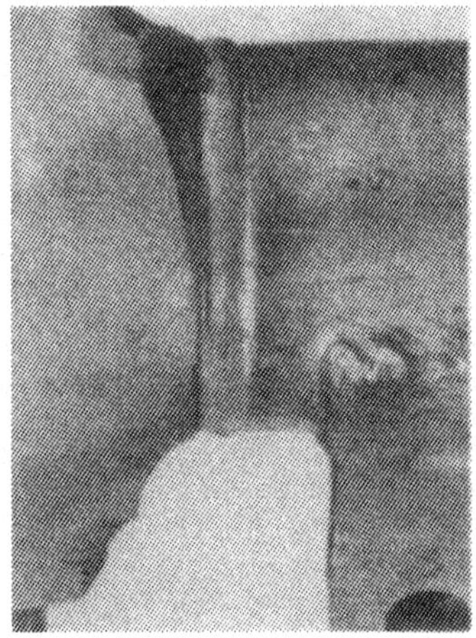

Fig. 266 One of the radial circumferential welds shown in Fig. 265 between the forged cheek (left) and the central tubular brace (right).

Fig. 267 Electron beam welded pre-combustion chamber for a diesel motor, ca 80 mm long un-machined.

with no sharp undercutting (Fig. 266). Including an induction pre-heating cycle, the entire welding process takes less than 2 min [148]. In the pre-combustion chambers of diesel engines dissimilar metals need to be joined (Fig. 267). The upper part is a high strength nickel alloy whilst the lower is steel X10 Cr 13. Pre-combustion chambers are welded in large numbers in

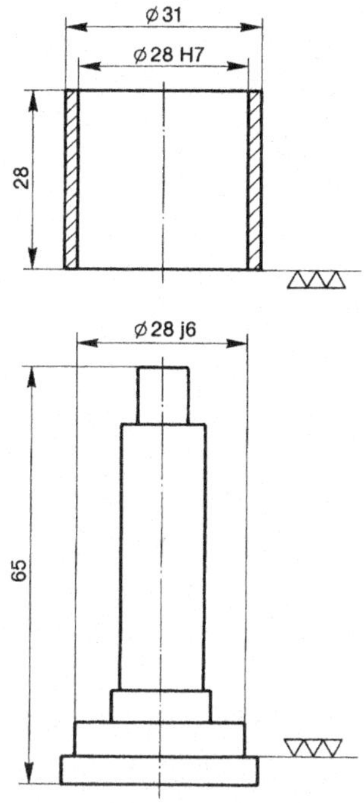

Fig. 268 Cross section through a shock absorber piston made of nitriding steel.

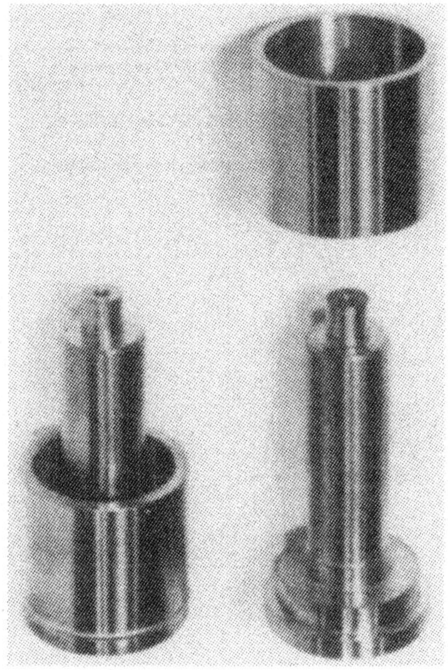

Fig. 269 An electron beam welded shock absorber piston before assembly.

continuous welding machines suitably equipped with pressure stage locking systems (Fig. 232).

10.3 Machine construction

A particularly interesting spectrum of different electron beam welding tasks is found in machine construction. The following examples are only a few of the many and varied uses of the process to weld both small and large workpieces economically.

The shock absorber pistons shown in Fig. 268 and 269 are made of an untreated nitriding steel. The main body of the piston and its shroud are fitted together with the tolerance H7/j6 which corresponds to a sliding fit. This is of no real consequence, however, as far as welding of the radial circumferential seam is concerned.

Thirty such pistons are welded in a single evacuating sequence using a multi-component rotating jig. To machine the part in one piece from solid material would be extremely costly.

The shell of a highly loaded plain bearing made of X22 CrNi 17 incorporates two cooling channels (Fig. 270). After machining, a U-shaped ring welded into place with two seams is used to seal each of the channels. After welding, the

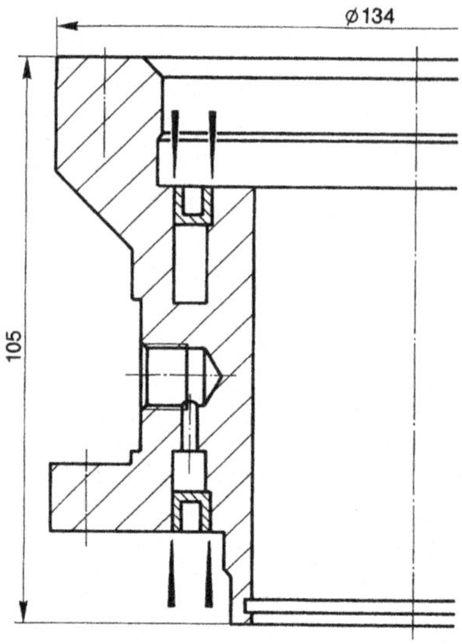

Fig. 270 A bearing housing with cooling channels. Material: X22 CrNi 17.

Fig. 271 An electron beam welded overlapped butt on a cam.

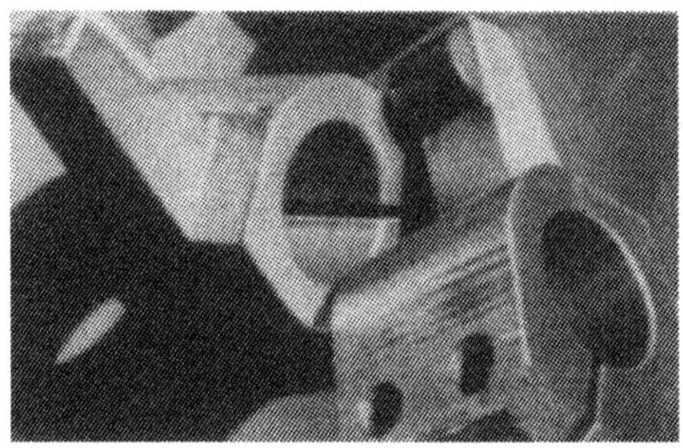

Fig. 272 A lever arm for a textile machine. Material: lever arm C 60; contact body 100 Cr 6.

shell is annealed under vacuum at 580 °C for 2 hours to reduce the hardness of the weld and heat affected zones.

As can be seen from Fig. 271, an overlap weld joins the 15 CrNi 6 lug to the G-42 CrMo 4 clamp. To increase the cross section of the joint the lap weld is executed in a zig-zag form by weaving the beam from side to side.

Figure 272 shows the contact body of a lever arm for a textile machine made from 100 Cr 6, an alloy normally considered unsuitable for welding. The contact body is welded to the lever arm made of C 60 steel by two 1 mm deep seams running along the face of the semi-cylindrical body.

Figure 273 shows an example of joining larger machine components, namely composite spun cast rolls for rolling mills (patent applied for).

The rolls have an outer diameter of 400 mm and above, a length of 1–3.5 m and weigh 0.6–5 t. The centre part of the roll is made up of two layers of

Fig. 273 Electron beam welded rolls made of spun cast composite material.

Fig. 274 Cross section through one of the spun cast rolls shown in Fig. 273. Left: The bearing mortice of 42 CrMo 4 4; Top right: Ferritic chromium steel; Centre right: 13 CrMo 4 4. (1:1; Reproduced 0.4×).

material spun cast together. The outer layer is a wear and corrosion resistant ferritic chromium steel whilst the inner layer is of high temperature 13 CrMo 4 4. The bond between the two layers of metal cannot be compared with a weld, being more similar to a brazed joint. The two bearing mortice keys are finish machined to very close tolerances. At the point of welding all three of the component materials are welded together to a depth of 60 mm (Fig. 274). The bonding of the spun cast layers and the finish machined mortice key require that a welding process be used which will have very little metallurgical effect and cause the smallest possible amount of distortion. This is only possible using electron beam welding. The cross section shown in Fig. 274 also shows how the base of the fusion zone, together with possible root cavities, has been shifted into a region of the workpiece which is insignificant as far as transfer of the dynamic forces generated during rolling is concerned.

10.4 Tools

Tools used in machining and cutting operations are subject to the most diverse operating stresses. At the cutting edge the tool should be hard and wear resistant whilst the carrier material is expected to be tough and have the ability to deform. For this reason cutting tools are subjected either to various different localised heat treatments or are made from a number of different materials. Joining of dissimilar materials is thus an important feature in electron beam welding of cutting tools. Figure 275 shows a segment of a circular saw blade, the teeth of which are made of HSS high speed steel and were previously rivetted to the blade which is made of C 60 steel. Using electron beam welding, a far more economical and durable bond between the two components is achieved. The weld seam is quite straight, such that a number of blade sections can be welded in a single evacuating sequence using only very simple jigs.

Fig. 275 Segment of a circular saw blade with electron beam welded teeth made of high speed steel.

Fig. 276 A bimetallic bandsaw blade ca 13 mm wide ($\frac{1}{2}''$) and 0.6 mm thick, b = 0.2 mm, v_s = 18 m · min^{-1}. Top: Electron beam welded; Bottom: Finish machined.

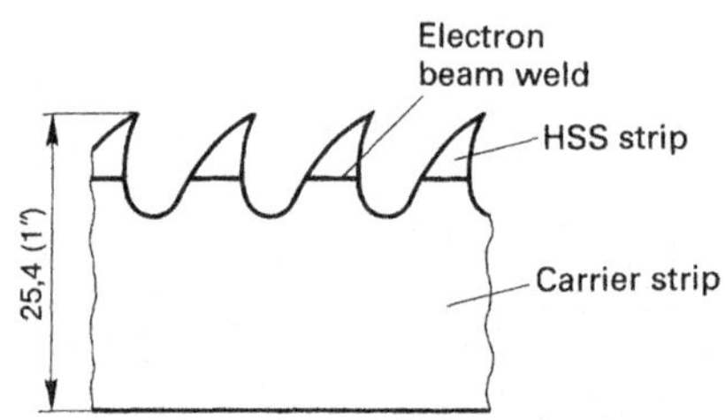

Fig. 277 Tips of the cutting teeth made of HSS-material with the base of the teeth in the carrier strip made of ca 25 mm (1") wide bimetallic bandsaw material.

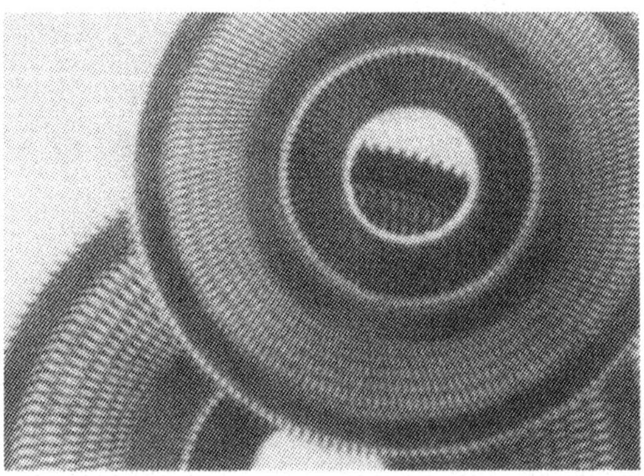

Fig. 278 Copper windings for disc armature motors the contact points of which are electron beam spot welded.

A further development of this possible use led to electron beam welding of bi-metallic bandsaw material, as has already been mentioned in section 8.7 (Fig. 276). An HSS strip, also referred to as flat wire, together with the carrier strip, is passed through vacuum seals into an electron beam welding machine in which it is continuously welded. After heat treatment the teeth are ground and set such that the tips of the teeth mainly consist of the HSS material whilst the gaps between the teeth are in the more durable carrier material (Fig. 277).

Fig. 279 Commutator with electron beam welded contact flags.

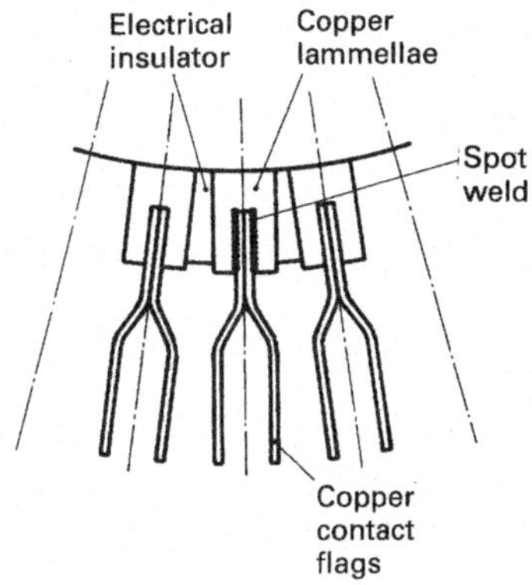

Fig. 280 The welded joints shown in Fig. 279.

10.5 Electric motor construction

In construction of electric motors, electron beam welding is used mainly for precision welding of electrical contacts. The copper windings of the disc armature motor shown in Fig. 278 are made up of two printed circuit boards which are welded together around their inner and outer circumferences. Welding is carried out with the boards tilted at an angle to the electron beam, to achieve a defect free joint between the wires in spite of the very small volume of the weld pool. In Fig. 279, the contacts are to be joined to the commutator strips. To achieve an adequate joint cross section while avoiding damage to the insulation material, the seams are welded point for point around the circumference of the commutator (Fig. 280).

10.6 Equipment construction

The base of the heat exchanger shown in Fig. 281 contains numerous holes to accept the tube bundles. With the aid of a CNC rotating arrangement each hole is positioned beneath the electron beam for welding and the circumferential seam welded with a circular beam oscillating motion. Welding of the entire base is carried out fully automatically. The vacuum proof instrument cable channel shown in Fig. 282 is also electron beam welded to preserve the integrity of the insulation by minimising the overall heat input.

Figure 283 shows a further example of precision welding in the housing of an overload diaphragm for a measuring instrument. The ring is joined to the disc from the inside with a T butt weld which is achieved by welding with an inclined rotational axis. The small wall thickness of the component parts requires that they are very accurately fitted to one another before welding.

Fig. 281 Heat exchanger with welded tube bundle. Wall thickness of tubes 1 mm, tube diameter 4 mm.

Fig. 282 Electron beam welded vacuum proof instrument cable channel.

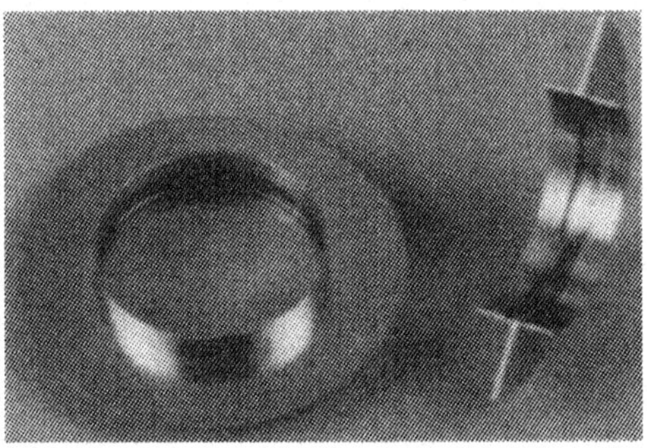

Fig. 283 Part of an overload safety diaphragm for a measuring instrument. Wall thickness 0.5–0.8 mm.

10.7 Medical technology

When welding the casings for heart pacemakers it is also very important that the highly sensitive electronic components are not damaged (Fig. 284). Here, highly accurate movement of the workpiece is required as well as careful preparation of the joint to be welded.

Figure 285 shows the flexible drive shaft of a surgical drill to which the cutter is attached by a circumferential weld seam. Because of the high degree of reliability required, electron beam welding has proven ideally suited to this task.

10.8 Economic considerations

Initially electron beam welding was almost the only possible means of fusion joining workpieces made of special metals such as molybdenum, tantalum, zirconium, etc, and thus the question of its economic viability was of secondary importance. Today, electron beam welding is used for joining all weldable materials, and is in strong technical and economic competition with other

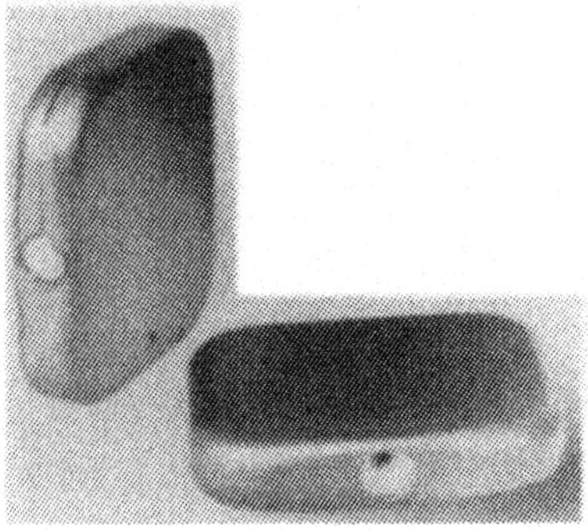

Fig. 284 Electron beam welded casing for a heart pacemaker.

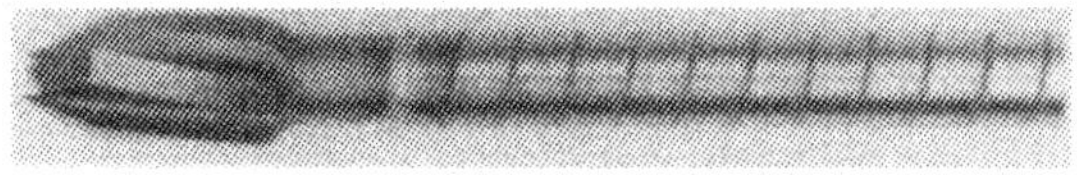

Fig. 285 Surgical drill with electron beam welded cutting head. (1:1; Reproduced 0.8×).

methods of joining materials, both for individual workpieces and in mass production. If the capital costs of electron beam welding equipment are high, the process must nevertheless have considerable advantages as it is used in such a large number of different branches of industry. The ability to weld without filler materials or other consumables, together with the very high speeds reached will not in most cases be sufficient grounds for such an economic justification. For this reason it is also incorrect to evaluate the process, as is done with other methods of welding, in terms of the characteristic cost per metre of weld run [149].

Typical costs for electron beam welding a given workpiece are made up of the following main factors:

- Preparation of the workpiece
  Even thick workpieces can be prepared with a cost-saving I butt preparation and do not need to be cut to a V, X or U shape as with other methods of welding. Careful machining of the faces to be joined is necessary, however, because the filler material which would otherwise fill the larger weld gaps is not used. Cleaning of the workpieces before welding should be the same as with other welding processes.
- Welding jigs
  This is also the same as with other methods of welding, but with the saving that no backing strips are used.
- Production time
  This is the basis for calculating various costs and comprises the time required to load the workpiece in the jig, to evacuate the working chamber, to carry out the welding, to vent the working chamber and to unload the workpiece. When welding with more than one workpiece in the working chamber at a time, the evacuating time is reduced proportionally.
- Personnel costs
  The costs of the operating personnel and any other assistance are made up of the time taken to weld a single component plus certain other

additional times (starting the welding machine at the beginning of the weld run, etc), and expressed as a cost per unit workpiece.

- Servicing and repair costs
  These costs are essentially made up of the cost of cleaning the working chamber and weld positioning equipment, together with those for servicing the vacuum system including pump oil and worn components, etc.
- Costs of consumables
  These would be the costs for the cathodes used, the electricity, cooling water, compressed air and, if relevant, the liquid nitrogen used to operate a cryogenic pump.
- Amortisation
  The cost of welding a single workpiece should also include a contribution to the capital amortisation of the welding machine.

A number of different savings must be set against these costs, resulting from a comparison with the cost of producing the component using other methods of manufacturing, including machining from solid material. The main cost savings are:

- High welding speed
  Welding any thickness of workpiece in one pass and at comparatively high speeds.
- Low distortion
  Welding accurately dimensioned workpieces whilst maintaining close tolerances. This offers many advantages in construction and manufacturing by being able to fabricate workpieces from a number of economically produced individual component parts. It is these advantages which generally permit the greatest cost savings to be achieved.
- No filler or additional materials
  In electron beam welding no additional materials such as shielding gas, powder or flux are used, and as a rule no filler material is added.
- No post-weld cleaning
  There is no slag, oxidation or strongly adhering spatter, etc, produced, which would otherwise require removing. When welding is carried out in a vacuum, the welds seams retain a metallically clean and bright finish.
- Weight saving
  The low 'spatial requirement' of the electron beam compared with for example the electric arc, or cutting tools when machining from solid material, is a particularly notable advantage in saving of weight and material costs.
- Welding of combinations of dissimilar materials
  The narrow fusion and heat affected zones produced permit a large number of combinations of different materials in almost any geometry of workpiece to be welded. This has numerous advantages in being

able to select the most suitable design of component in respect of its loading and function.

This listing and description of the different factors affecting both the costs and cost savings involved in electron beam welding can only act as a guide in any assessment of the economic viability of the process for a given application. In each individual case, the user must determine which costs are applicable to the particular operation, and which of the advantages of the electron beam welded component are of greatest importance and how these advantages are to be estimated financially. Depending upon the particular branch of industry, the competition involved and the in-house conditions and environment, very different results can be obtained from such an assessment even for very similar examples of welding. It is thus impossible to quote numerical data here since they would be valid only for a given component in one particular operation, would not be transferable and would soon become out of date because of the rapid technical developments at present taking place. As mentioned at the beginning of this book, however, the large number of electron beam welding machines in use in industry must prove that, in spite of the significant investment costs involved, the process is nevertheless characterised by a high economic viability.

11 Personnel qualifications and machine testing

11.1 General

For each product, quality requirements are set which depend not only upon the function of the product but also upon the possible consequences should it fail in service. If the quality of a product is to be influenced, then the factors which affect it must be known. Table 9 summarises factors which directly or indirectly affect the quality of electron beam welded components. This summary is also valid technically for components manufactured using other mechanised welding processes, although in these cases other specialist terminology would normally be more applicable. The following sections deal with qualification of personnel and with testing of electron beam welding machines. The material, workpiece, and evaluation of defects in the weld have already been discussed in previous sections.

11.2 Personnel qualification

In all mechanised welding processes, training and testing of the operating personnel has so far remained very much in the background. Even today, it is left to the users of electron beam welding machines themselves, or to the manufacturers of such equipment, to introduce personnel to this method of welding and to train them to operate the machines used.

Whilst the welder certification often required in manual welding is a useful basis for electron beam welding, it does not meet the requirements for the operating personnel. With the exception of a single DVS information sheet [150] there is no documentation regarding entry level requirements, course content or method of examining such a course of instruction.

It is possible that the DVS guidelines [151] for training to become a Laser Beam Specialist could in future be adapted to the requirements of electron beam welding. The entry level requirement for this is a completed course of study in a metal working or electro-technical profession together with an in-depth knowledge of welding and welding equipment and the behaviour of materials on welding, as well as of CNC systems as required in the training of welding engineers or for the operation of CNC machine tools. The in-depth knowledge could also be gained from a suitable course of preparatory study.

In Germany, it is planned to divide the main course of study for an electron beam welding engineering course into the following sections:

Table 9 Factors affecting the quality of electron beam welds

1. Direct Factors

1.1 The material
- Analysis, method of production
- Heat treatment
- Surface treatment
- Magnetism
- Strength and corrosion properties

1.2 The workpiece
- Design shape
- Weld preparation
- Cleanliness
- Internal stresses, deformation

1.3 The welding machine
- Parameter optimisation
- Constancy of the settings and positional stability
- Accuracy of the workpiece positioning systems
- Beam guidance system
- Machine control system
- Welding jigs
- Servicing and maintenance

2. Indirect Factors

2.1 The operating personnel
- Specialist knowledge
- Reliability
- Commitment
- Demands
- Responsibility

2.2 Weld testing
- Identification and determination of changes
- Evaluation of changes

2.3 Company conditions
- Company organisation
- Work planning
- Training
- Exchange of information

- The basics of the electron beam welding process
  - Beam generation
  - Beam diagnostics
  - The welding process
  - The weldability of metals
  - Workpiece preparation
  - Vacuum technology
  - Beam and machine control systems
  - Radiation protection
- Machine operation
  - Changing of cathodes
  - Optimising the set welding parameters
  - Working with control systems

Evaluating welds
Servicing
Welding documentation and procedures
- Practical welding
  Welding of practical examples
- Final examination
  Theory
  Practical welding

The duration of the preparatory course of study would need to be 2 weeks, and for the main course of study 4 weeks. On completing the course the electron beam welding engineer should be able to undertake special tasks independently and with the minimum of supervision.

It is repeated, however, that this is only a model of a course based on that at present run for the Laser Beam Specialist. It is interesting that major users of electron beam welding machines already train their own personnel internally in a similar way, but of course in each case tailored more specifically to the particular needs of the individual companies concerned.

11.3 Machine testing

11.3.1 Basics, acceptance criteria

In contrast to personnel training, for testing electron beam welding machines there is a detailed DIN standard which has been well proven in practice and which can be adapted, namely DIN 32 505. This standard is based upon the principle that constant and reproducible quality in the manufacture of high grade welded joints is assured if, amongst other factors, the required welding parameters set remain constant and reproducible within certain given tolerances over the period of operation.

In the field of acceptance testing, which is mainly involved with commissioning of such machines at the users' works, the short and long term variation of the main welding parameters (e.g. accelerating voltage, beam current, lens current and welding speed) are measured and evaluated according to this standard. Furthermore, the accuracy of the positioning equipment and the positional stability of the electron beam spot are also measured to see if they meet their respective requirements (Fig. 286).

DIN 32 505 Part 1 describes certain requirements in respect of the mains power supply, radiation protection and the positioning of the installation itself, which must be addressed before acceptance testing is begun. Finally, the working range of the electron beam welding machine must be defined. Here, the ranges of the parameters U_{Bmax}, U_{Bmin}, I_{Smax}, I_{Smin}, the limits of the working distance, the maximum movements of the positioning equipment and the maximum safe weight of workpiece and welding jig are set.

Whilst within this working range the limits of variation must be strictly adhered to, but during acceptance testing only a restricted number of previously specified settings are measured.

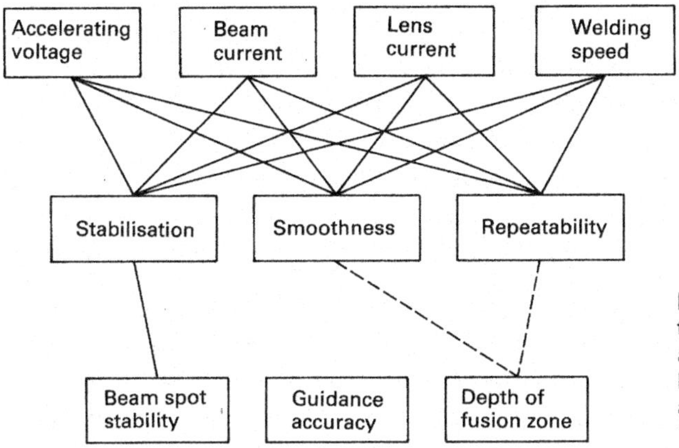

Fig. 286 Relationship between the factors involved in acceptance testing of electron beam welding machines according to DIN 32 505 Parts 1–6.

Table 10 Maximum acceptable variation in electron beam welding parameters (from DIN 32 505 Part 1)

Parameter		Tolerance limit
Accelerating voltage		
Residual ripple		2% (peak to peak)
Stability		±1%
Reproducibility		±1%
Beam current		
Residual ripple		5% (peak to peak)
Stability		±1%
Reproducibility		±1%
Lens current		
Residual ripple		0.5% (peak to peak)
Stability		±0.5%
Reproducibility		±0.5%
Welding speed		
Short term constancy	υ_{max}	±2%
	$0.1\upsilon_{max}$	±5%
Long term constancy		±1%
Reproducibility		±1%
Accuracy of positioning		$\pm 0.05\,mm \cdot m^{-1}$
Beam spot constancy		$\pm 0.1\,mm \cdot 300\,mm^{-1}$

Table 10 summarises the maximum acceptable variations for the various parameters. The term residual ripple means the variations which occur from a mean value over a milli-second timespan, which can only be measured using an oscilloscope. The cause of these variations is mainly incomplete rectification of the alternating voltage from the mains power supply into a direct voltage. Short term variations in welding speed are caused by vibration and play in the mechanical parts of the positioning equipment, as well as by dirt and wear in the working table or rotating mechanism. It is thus important that from time to time in service, the stability of speed of this equipment is checked.

In measuring the stability of the system, long term variations are determined which could be the result of thermal effects in the electrical control circuits. The limits of stability are of particular importance whenever welding times are long, such as for example in welding large diameter circumferential seams. In DIN 32 505 Part 1, an operating period of 30 min is specified for the stability of the system, and of three times the maximum welding time for the positioning systems.

To test the reproducibility of the system, the variations in the different parameters are checked after repeatedly switching the system on and off. During the test no adjustment or correction of the system is permitted.

The variations thus determined are of particular significance in automatic electron beam welding of serial components.

If all the variations measured are within the limits shown in Table 10, then the suitability of the machine for electron beam welding is demonstrated according to the standard. However, for the majority of possible uses of the electron beam welding process other additional criteria will also be used to judge a machine's suitability. Thus in practice the scope of testing applied would normally be widened or restricted, as required by discussion between the user and manufacturer.

If for example the machine is to be used to weld I butt seams with centering lips whilst maintaining a closely toleranced depth of fusion zone, then variations in depth of the fusion zone would need to be included in the scope of testing. In practice this is done by welding three seams at different times and measuring the depth of their fusion zones on longitudinal macrosections through the welds. In this case the settings, material and geometry of the workpiece used should be chosen according to the actual use to which the machine is to be put.

Retests are normally only required when changes have been made to the electron beam welding machine which could affect its suitability for use (e.g. rebuilding, repair, replacement of parts, dismantling, change of location, etc). In addition, the fitness for purpose of the machine will also be documented through continual reports that the welding being carried out is of satisfactory quality.

11.3.2 Measuring the accelerating voltage

The voltages generally used in electron beam welding, in particular those at the output side of the high voltage power supply and at the termination of the high voltage cable in the beam production chamber, cannot be measured safely without special equipment and instrumentation. Attempts to measure a possible variation in accelerating voltage U_B indirectly by measuring the fusion track of a deflected beam (Fig. 287) generally fail because of the inherent inaccuracy of this method. Accelerating voltages should thus only be measured directly and using special equipment. Normally, however, only the manufacturers of welding machines or high voltage installations will have this equipment capability, and thus it is normally best to have such measurements made

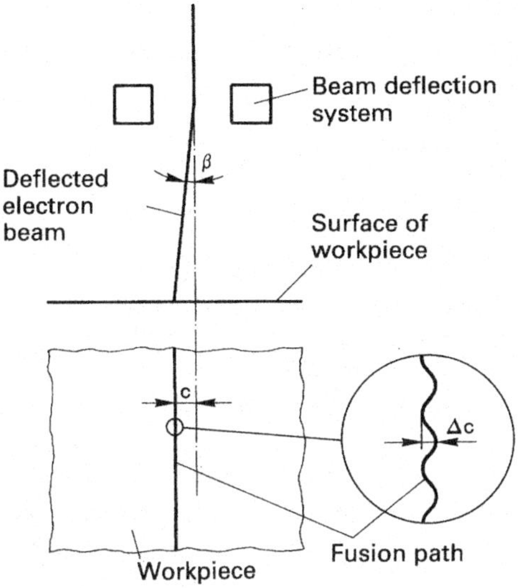

Fig. 287 A method of measuring the smoothness of the accelerating voltage U_B at constant beam deflection.

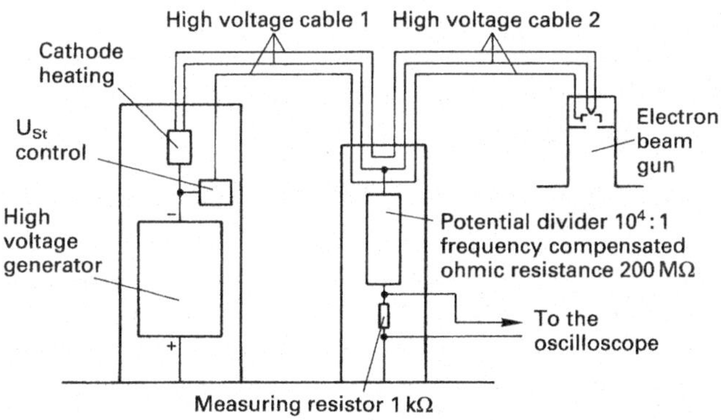

Fig. 288 A system for measuring U_B using a potential divider in accordance with DIN 32 505 Part 2.

at the manufacturer's own works. In addition to the safety aspect, the equipment must also be capable of measuring the accelerating voltages over the entire range of setting of the beam current, or in other words at normal welding settings.

In measuring U_B as shown in Fig. 288, a voltage divider is inserted between the high voltage supply and the beam gun. Like the high voltage conducting elements of the supply, this is also housed in a suitably sized oil-insulated casing and is connected to the electron beam welding machine by the normal type of high voltage cable. The electrical values of the voltage divider are temperature dependent and thus the voltage divider must be calibrated for a particular insulating oil temperature, which must be held constant whilst the measurements are made. The dimensioning of the frequency compensation depends upon the type of construction of the high voltage supply and aims to make the voltage divider act purely as an ohmic resistance.

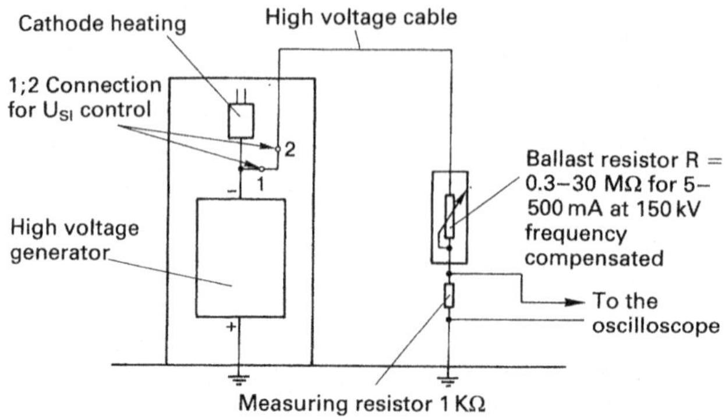

Fig. 289 A system for measuring U_B using a ballast resistor in accordance with DIN 32 505 Part 2.

Instead of the voltage divider, a stepping resistance may also be used (Fig. 289). This replaces the beam gun and also allows the accelerating voltage to be measured at all specified beam currents and operating times. Here also, temperature dependency and frequency compensation must be taken into account.

In both cases the voltage difference is measured across a measuring resistance and, with the aid of an oscilloscope or X/Y recorder, the smoothness, stability and repeatability are determined and compared with the permissible variations from the specified limits shown in Table 10.

11.3.3 Measuring the beam current

The beam current is also measured by restriction in a solid current conductor using a special piece of equipment. For this a Faraday cage is used, which if possible captures all the electrons entering the working chamber and conducts them to earth as a current through a defined resistance (Fig. 290). The beam impingement current I_A thus determined, because of scattering effects, is slightly smaller than the actual electron current emitted by the cathode which is measured in the high voltage supply as the beam current I_S and is registered on the operating console. However, this has no effect on the evaluation of stability or reproducibility of the current.

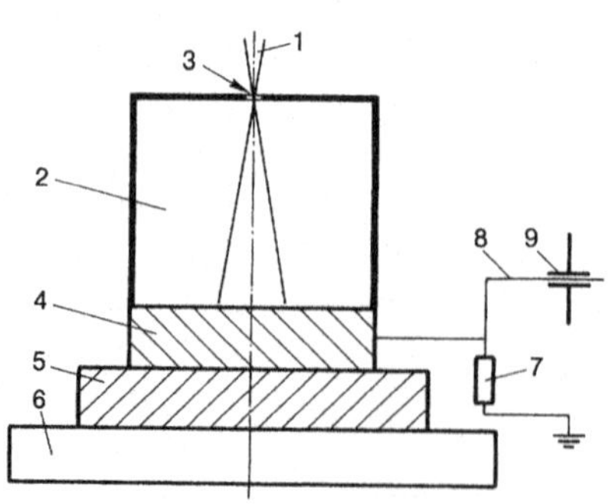

Fig. 290 A Faraday cage for measuring I_S in accordance with DIN 32 505 Part 3. 1 Electron beam; 2 Faraday cage; 3 Aperture; 4 Base of cage; 5 Electrical insulation; 6 Work table; 7 Measuring resistance; 8 Screened cable to the oscilloscope; 9 Vacuum feed through the working chamber.

DIN 32 505 Part 3 gives details of different Faraday cups which, depending upon the accelerating voltage U_B and the beam power P, are dimensioned according to the thermal effects to which they will be subjected during the course of measurement. The cups are formed such that they permit the entry of the focused electron beam and largely prevent the escape of primary and secondary electrons. To measure the beam current the cup is positioned, electrically insulated, in the working chamber and is connected to earth through a resistance. A vacuum sealed feed through the chamber permits the voltage difference across the resistance to be measured using an oscilloscope and the beam current thus to be measured and evaluated according to the different requirements.

It should be realised that at high beam currents large areas of the bottom of the cup will be melted, and as a result errors in measurement could occur. This difficulty can be minimised by using short measuring times separated by long periods of cooling, if necessary at the same time venting the working chamber.

It is recommended that the beam current is measured in this way at regular intervals, including a comparison of I_A with I_S, in order to build up a picture of the overall stability and reproducibility of the most important beam parameters.

11.3.4 Measuring the welding speed

Whilst measuring and setting the required values for the electrical parameters U_B, I_S and I_L present no particular difficulty, for a long time electron beam welding machines had no satisfactory instruments for measuring the equally important welding speed. For measuring this 'mechanical' parameter there was normally only a potentiometer used for controlling the drive motor. Measurement of the actual speed of the working table or rotating mechanism was not possible.

The situation changed only with the introduction of incremental measuring transducers, which are able to transform a translation or rotational movement directly into a proportional electrical signal.

For acceptance tests encoders have proven useful: together with frequency encoders, they are able to produce a voltage signal directly proportional to the speed measured. In the simplest case a hard rubber wheel is pressed against the moving working table, workpiece or rotating arrangement, this driving the shaft of an encoder (Fig. 291 and 292). A stroboscopic disc fixed to the shaft is sensed by an optical transducer such that, on rotation, variations in brightness are turned into a series of electrical impulses by a series of photocells. The impulse frequency and the resulting comparative voltage are thus a measure of the speed of the moving object. During measurement the rubber wheel is sprung against the contact surface of the component to be measured (Fig. 291 and 292). Any possible skidding of the friction wheel is immediately apparent during the measuring process and can normally be prevented by increasing the contact pressure. On PLC and CNC machines the signals can be obtained from an encoder connected directly to the drive spindle, see Fig. 198.

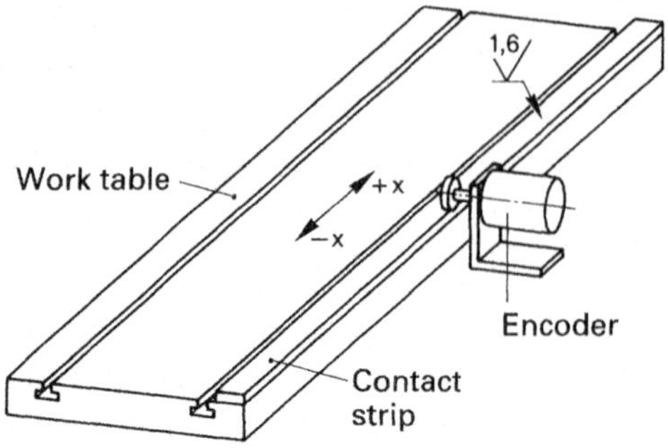

Fig. 291 System for measuring the constancy of speed of translational movements in accordance with DIN 32 505 Part 4. Example: for movements in the +X and −X directions without supporting the weight of the workpiece.

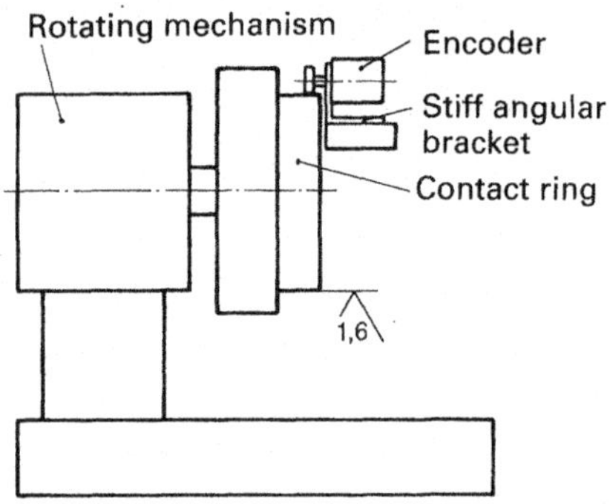

Fig. 292 System for measuring the constancy of speed of rotational movements in accordance with DIN 32 505 Part 4. Example: for movements in the +A and −A directions without supporting the weight of the workpiece.

As described in detail in DIN 32 505 Part 4, measurements should not be carried out solely at different operating speeds, but also on the actual workpiece with the positioning mechanism under different applied loads.

For testing rotating positioners with horizontal axes, this means that the tilting moment and any possible non-uniform distribution of the weight of the workpiece must also be taken into account (Fig. 293).

The actual speed must be measured directly on the working table or rotating mechanism in order also to measure any possible irregularities in the transfer of movement, e.g. as might occur when using a gear or spindle drive. Measurement of a simple 'tachometer voltage' is not considered to be adequate to check stability of speed.

11.3.5 Measuring the tracking accuracy

The working table and rotating mechanism are expected to produce movements of the workpiece of a certain accuracy. The term tracking accuracy means the maximum variation occurring at right angles to a given change in movement in the X, Y and Z directions, or flat and eccentrically in the case of rotated workpieces. This also includes any variation in the position of the focal point of the beam from its set position, as far as this might be caused by the positioning mechanism.

DIN 32 505 Part 5 includes advice on instruments and instructions on how

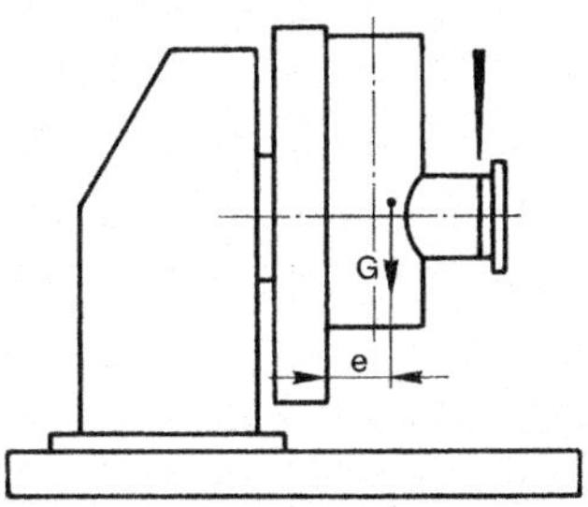

Fig. 293 Non-uniform distribution of the mass of the workpiece, significant when measuring the constancy of speed and accuracy of movement of the workpiece positioning system. G·e = Moment of tilt.

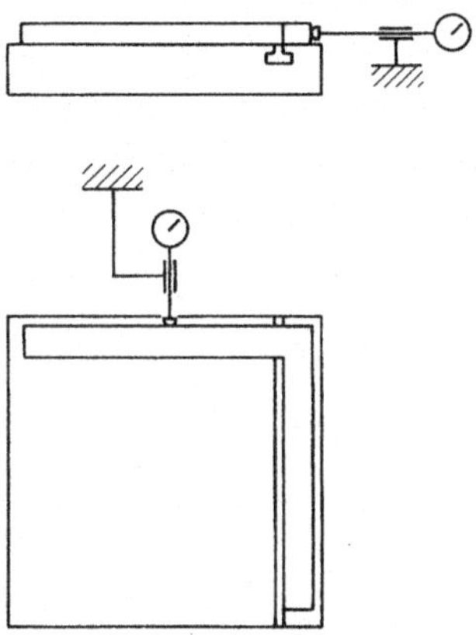

Fig. 294 A method of measuring the accuracy (top) and squareness (bottom) of movement of the work table in accordance with DIN 32 505 Part 5.

to measure accuracy of movement. With the sheer number and diversity of positioning arrangements and welding machines used, the standard cannot cover the complete range of measurements required. However, in all cases measurements of the linear and transverse tracking accuracy of the working table (Fig. 294), and of the flatness and eccentricity of rotating equipment should be made. Here also different and irregular distributions of workpiece weights should be taken into account as shown in Fig. 293.

11.3.6 Measuring the stability of the beam spot

The beam spot position refers to the point of impingement of the electron beam on the surface of the workpiece. A constant beam spot position is important in all types of electron beam welding to prevent fusion defects. This is particularly important with long welding times and in the case of electron beam welding machines not equipped with optical viewing systems. Figure 295 shows the design of an instrument used to measure the stability of beam spot position. The measuring sequence consists of the following steps:

- Adjust the electron beam to weld a point on workpiece A.
- Bring the Y and X co-ordinates of workpiece B into line with the crosswires in the eyepiece of the optical viewing system.

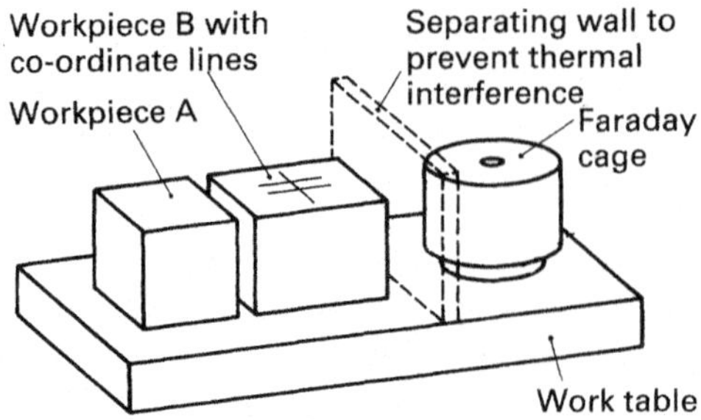

Fig. 295 A method of measuring the constancy of size of the focal spot in accordance with DIN 32 505 Part 6.

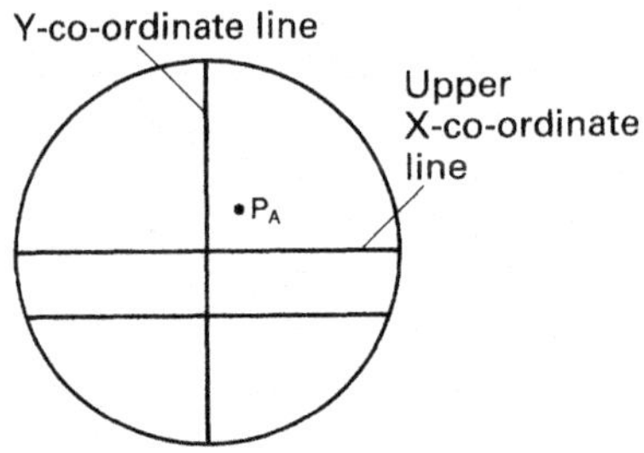

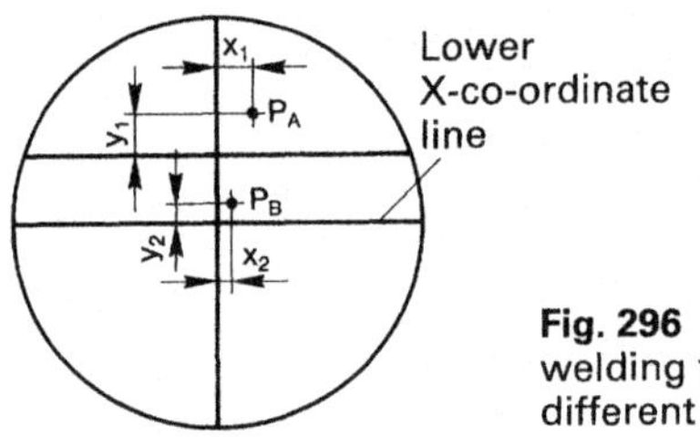

Fig. 296 Field of view in the optical inspection system after welding the first point P_A and the second point P_B, showing the different measurements as in DIN 32 505 Part 6.

- Weld the first point P_A (Fig. 296, top).
- Position a Faraday cup on the axis of the beam and turn on the electron beam at 10% of P_{max} for the required operating time.
- Adjust the electron beam to weld another point on the workpiece A.
- Bring the Y and X co-ordinates of workpiece B into line with the crosswires in the eyepiece of the optical viewing system.
- Weld the second point P_B (Fig. 296, bottom).
- Measure the co-ordinates of the welded points and calculate any difference between them. The values determined should then be calculated proportionally to $A_F = 300\,mm$ in case during welding of the points a different focal distance was used.

By welding two separate points even small variations in the position of the beam spot can be measured. Of course, between the individual steps of the measuring process no alterations, and in particular no adjustments, may be made to the welding machine.

As far as welding machines not equipped with optical viewing systems are concerned, DIN 32 505 Part 6 gives only the basic advice that the method of measurement to be used to determine the stability of the beam spot depends upon the construction and method of operation of the machine. Instead of points, blind concentric circumferential seams of different diameters are welded. In such a case, the mechanical arrangements for positioning the workpiece must be tested separately.

11.4 Radiation protection

In Chapter 3 it has already been explained that on penetrating a material electrons produce X-rays. At accelerating voltages of $U_B \leqslant 60\,kV$, the energy of the X-rays produced is very small and is reduced, without need for additional shielding, to a value below that of the maximum permissible local dose rate by the casing of the electron gun and the walls of the working chamber. Any viewing windows should, however, be made of leaded glass. At accelerating voltages of $U_B > 60$ kV the electron gun and working chamber need to be additionally shielded with lead cladding. On no account should the electron beam be turned on if the lead shielding has been damaged or partially removed, or the leaded glass window replaced by an aluminium or steel cover.

According to the current German radiation protection regulations (X-rays) (of 8 January 1987), see section 12.3, electron beam welding machines are classified as X-ray emitting sources and require a licence according to which the operator of the welding machine is required:

- prior to commissioning, to obtain the permission and required licence from the authority responsible (e.g. the Department of Industry);
- to appoint a member of staff responsible for radiation protection and if necessary a radiation protection supervisor;
- to train the operating personnel suitably in correct use of the electron beam welding machine and in the radiation protection requirements, this training to be repeated annually;
- to fix a sign to the machine warning that modification or removal of any of the component parts of the machine relevant to its radiation protection would immediately invalidate the licence to operate the machine.

The radiation protection of the machine should be tested at least prior to its initial licensing by a qualified third party. To do this, amongst other tests, the defocused electron beam at power U_{Bmax} is aimed at a freestanding target, generally made of tungsten, suitably positioned in the working chamber.

The local dose rate is then measured at a distance of 10 cm from the outside surface of the walls of the working chamber and the electron gun using an X-ray detector. In the case of a professionally manufactured and correctly operated electron beam welding machine, the stipulated maximum permissible local dose rate will easily be complied with.

12 Standards and regulations

The following is a summary of the standards or guidelines currently applicable or under review which are significant to beam welding and cutting processes [152].

Correct as at: September 1992

12.1 DIN and EN standards

Number	Title	Edition	Comments
DIN 1910 Part 2	Schweißen; Schweißen von Metallen, Verfahren Welding; Welding of metals, process	08.77	EB = Electron Beam Welding
DIN 8563 Part 11	Sicherung der Güte von Schweißarbeiten; Elektronenstrahl und Laserstrahl Schweißverbindungen an Stahl, Bewertungsgruppen für Unregelmäßigkeiten Ensuring the quality of welding; electron beam and laser welded joints in steel; quality for irregularities	03.92	
DIN 29 595	Luft- und Raumfahrt; Schmelzgeschweißte metallische Bauteile, Technische Lieferbedingungen Aerospace; Fusion welded metallic components, technical specifications	04.83	
DIN 32 505 Part 1	Abnahmeprüfungen für Elektronenstrahl-Schweißmaschinen; Grundlagen, Abnahmebedingungen Acceptance inspection of electron beam welding machines; Principles and acceptance conditions	10.87	
DIN 32 505 Part 2	Abnahmeprüfungen für Elektronenstrahl-Schweißmaschinen; Messung der Beschleunigungsspannung Acceptance inspection of electron beam welding machines; Measurement of accelerating voltage	08.88	
DIN 32 505 Part 3	Abnahmeprüfungen für Elektronenstrahl-Schweißmaschinen; Messung des Strahlstromes Acceptance inspection of electron beam	10.87	

12.1 (*Cont.*)

Number	Title	Edition	Comments
	welding machines; Measurement of beam current		
DIN 32 505 Part 4	Abnahmeprüfungen für Elektronenstrahl-Schweißmaschinen; Messung der Schweißgeschwindigkeit Acceptance inspection of electron beam welding machines; Measurement of welding speed	10.87	
DIN 32 505 Part 5	Abnahmeprüfungen für Elektronenstrahl-Schweißmaschinen; Messung der Führungsgenauigkeit Acceptance inspection of electron beam welding machines; Measurement of accuracy of movement	04.89	
DIN 32 505 Part 6	Abnahmeprüfungen für Electronenstrahl-Schweißmaschinen; Messung der Flecklagekonstanz Acceptance inspection of electron beam welding machines; Measurement of welding point position tolerance	10.87	
DIN 32 511	Elektronenstrahl- und Laserstrahlschweißen; Begriffe für Verfahren und Geräte Electron beam and laser welding; Terms for processes and equipment	08.87	New version in preparation
DIN 65 118 Part 1	Luft- und Raumfahrt; Geschweißte metallische Bauteile, Angaben in Bauunterlagen Aerospace; Welded metallic components, details in construction documentation	03.92	
DIN 65 118 Part 2	Luft- und Raumfahrt; Geschweißte metallische Bauteile, Konstruktionshinweise Aerospace; Welded metallic components, construction advice	03.92	
DIN 65 171	Luft- und Raumfahrt; Abnahmeprüfungen für Vorrichtungen und Zusatzausrüstungen zum Schmelzschweißen, Führungs- und Bewegungsgenauigkeit Aerospace; acceptance testing for jigs and additional equipment for fusion welding, accuracy of tracking and movement	03.84	
EN 287 Part NN	Prüfen des Bedienpersonals für mechanisierte Schweißverfahren Testing of operating personnel for mechanised welding processes		In preparation
EN 288 Part NN	Anweisung zum Elektronen- und Laserstrahlschweißen		In preparation

Number	Title	Edition	Comments
	Instruction for electron beam and laser welding		
EN 288 Part NN	Verfahrensprüfung zum Elektronen- und Laserstrahlschweißen Procedure testing for electron beam and laser welding		In preparation
EN 729	Schweißtechnische Qualitätsicherungsysteme Quality assurance systems in welding		In preparation
EN . . .	Härtemessungen an Strahlschweißnähten Hardness testing of beam welds in steel		In preparation

12.2 DVS information sheets and guidelines

Number	Title	Edition	Comments
Guideline 1187	DVS-Lehrgang, Fachkraft für die Metallbearbeitung durch Laserstrahl DVS-Course, Trained specialists for laser beam metal-working	04.91	
Information sheet 2703	Elektronenstrahlschweißen in der Luft- und Raumfahrt Electron beam welding in the aerospace industry	08.75	
Information sheet 2708	Instandsetzung durch Schmelzschweißen im Luft- und Raumfahrzeugbau Remedial fusion welded repairs in aerospace vehicle construction	08.79	
Information sheet 2713	Schweißen von Titanwerkstoffen Welding of titanium based alloys	06.87	
Information sheet 2803	Elektronenstrahlschweißen in Elektronik und Feinwerktechnik Electron beam welding in electronics and instrument technology	12.74	
Information sheet 3201	Grundsätze für das Kontstruieren elektronenstrahlgeschweißter Bauteile Principles for the construction of electron beam welded components	11.86	
Information sheet 3202	Elektronenstrahl- und Maschinensteuerungen Electron beam- and machine control systems	06.88	
Information sheet 3203 Part 1	Qualitätssicherung von CO_2-Laserstrahl-schweißarbeiten; Verfahren und Laserstrahl-schweißanlagen Quality control of CO_2-Laser beam welding; Processes and beam welding machines	12.88	

Number	Title	Edition	Comments
Information sheet 3203 Part 2	Qualitätssicherung von CO_2-Laserstrahl-schweißarbeiten; Prüfen von Schweiß-parametern Quality control of CO_2-Laser beam welding; Testing of welding parameters	01.90	
Information sheet 3203 Part 3	Qualitätssicherung von CO_2-Laserstrahl-schweißarbeiten; Schweißneigung metallischer Werkstoffe Quality control of CO_2-Laser beam welding; Weldability of metallic materials	12.88	
Information sheet 3203 Part 4	Qualitätssicherung von CO_2-Laserstrahl-schweißarbeiten; Nahtvorbereitung und konstruktive Hinweise Quality control of CO_2-Laser beam welding; Weld preparation and advice on construction	06.88	
Information sheet 3204	Elektronenstrahl-Schweißeignung metallischer Werkstoffe Electron beam weldability of metallic materials	06.88	
Guideline 3205	Schutz vor Röntgenstrahlen an Elektronenstrahlmaschinen zur Materialbearbeitung Radiation protection from electron beam metal working machines	02.92	In draft

12.3 Other regulations

Number	Title	Edition	Comments
VdTÜV-Information sheet Welding Technology 1159	Verfahrensprüfung und Arbeitsprüfung für elektronenstrahlgeschweißte Bauteile Weld procedure and weld testing for electron beam welded components Weld procedure and weld testing for electron beam welded components	11.85	
RöV (Röntgenverordnung)	Verordnung über den Schutz vor Schäden durch Röntgen-Strahlen Laws governing the protection from radiographic radiation	01.87	BGBl. I. 114

13 Other methods of working materials with electron beams

13.1 General

The main aim of the electron beam welding process is to melt the contacting surfaces of two workpieces butted against one another to permit the melted material to flow together and solidify. There are nevertheless other phases which take part in the welding process. In melting the full thickness of the workpiece, some of the material is transformed into the vapour phase whilst another part of the workpiece is so slightly heated that it remains in the solid phase but may undergo metallurgical changes. The excellent control which can be exerted over the electron beam process will, however, also permit material to be heated without causing it either to melt or vaporise. This makes possible a whole series of new working processes for metals as well as for plastics and ceramics, (Fig. 297) [153]. In these applications the electron beam is used as a thermal nonwearing tool, either to modify the properties of, or to remove, material [154].

The main technical advantages of such processes are as follows:

- Transformation of the power of the beam into heat regardless of the optical properties and roughness of the surface of the workpiece;
- Numerous possibilities of processing large areas of the workpiece using inertia free beam guidance;
- High reproducibility of the beam parameters;
- Excellently suited to integration into series production.

13.2 Hardening

In electron beam hardening of steel the surface of the workpiece is heated locally until it transforms to austenite, which is then rapidly cooled by the adjacent unaffected material. As a result of the intimate material contact at the point of impingement of the electron beam and the high thermal conductivity of the material, extremely high cooling rates occur which result in high hardnesses being produced (Fig. 298). This technique is of course only suitable for transformation hardening steels with adequate carbon contents. The advantage of electron beam hardening lies in the possibility of treating points, lines or areas of such surfaces without metallurgically affecting other adjacent areas of the workpiece. In this way the hardened zones produced can be better adapted to suit the geometry, function and expected pattern of wear of the workpiece than for example is possible using induction hardening. In addition, with electron beam hardening no additional cooling medium such as water or

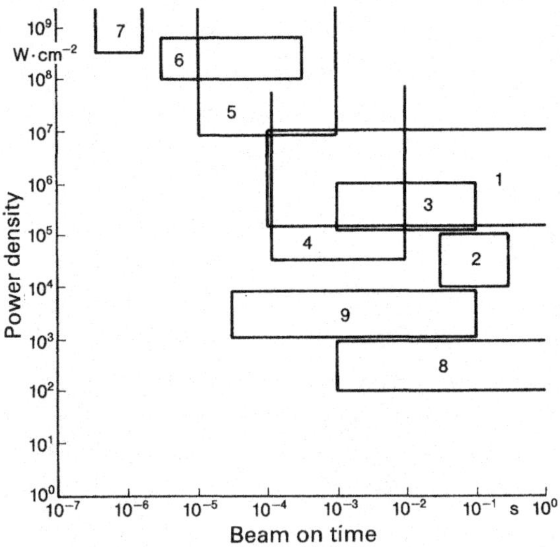

Fig. 297 Power density and working times of various electron beam processing techniques.
1 Welding; 2 Hardening; 3 Remelting; 4 Metal perforating; 5 Metal drilling; 6 Engraving; 7 Sublimating; 8 Polymerisation of plastics; 9 Perforation of plastics.

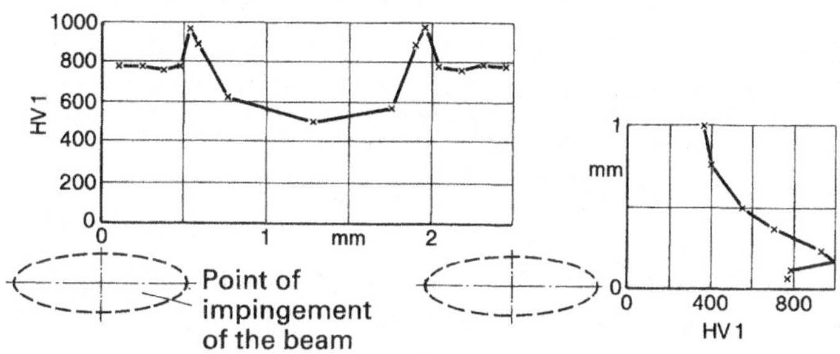

Fig. 298 Hardness values obtained when electron beam spot hardening 40 CrNiMo 8 3.

oil is required. The speed of electron beam hardening is, however, limited metallurgically. During the heating phase the iron lattice requires a certain minimum time (of the order of tenths of a second) to transform from the body centered cubic to the face centered cubic form at the austenitising temperature. Thus, increasing the beam power does not lead to accelerated transformation but only to an increase in temperature of the metal causing the surface to begin to melt.

This also explains the relatively shallow effect of electron beam hardening which is characteristically between 0.1–0.3 mm.

13.3 Re-melting

In this process the electron beam heats the surface of the workpiece, without the deep welding effect taking place, to a temperature slightly above that of the melting point of the material. In this case, the rapid cooling cycle affects the metallurgy of the material in a somewhat different way from hardening. Here, a fine grained microstructure is formed from the supersaturated solid solution,

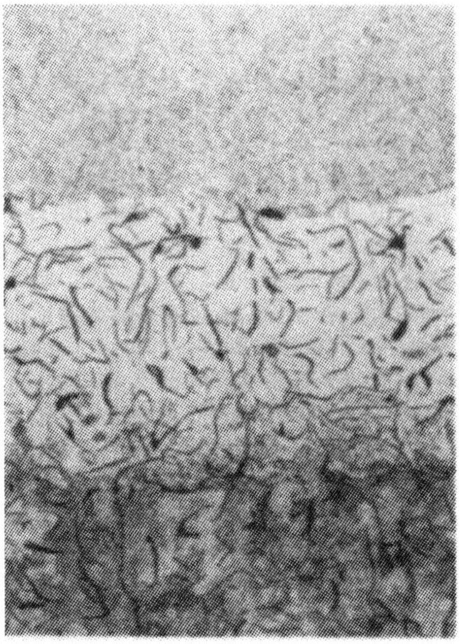

Fig. 299 Microstructure of remelted lamellar cast iron, from top to bottom: fusion zone; heat affected zone; parent metal. (100:1, Reproduced: 0.7×).

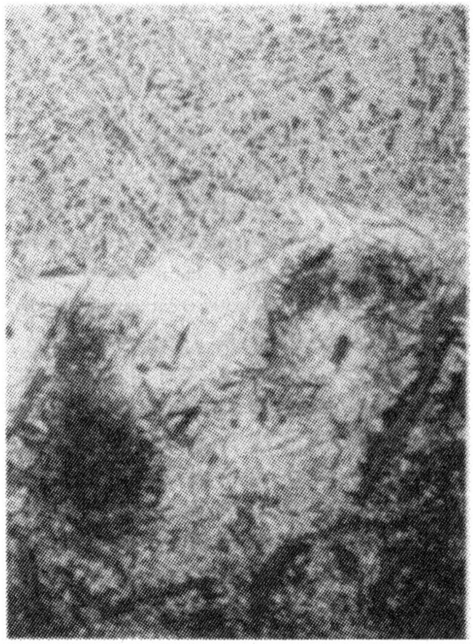

Fig. 300 Microstructure of remelted lamellar cast iron, transition from the fusion zone to the heat affected zone. (500:1, Reproduced: 0.7×).

which decreases segregation and transforms brittle intermetallic phases into ductile.

Only those materials whose microstructure is altered by changes in the rate of solidification are suitable for electron beam re-melting. These are generally materials of complex alloy composition. Thus, after re-melting, specimens of the high temperature steel X 20 CrMoV 12 1 exhibit an increase in hardness of from 230 to 550 HV0.3 together with an increase in fatigue strength under reversed bending conditions ($n = 2 \cdot 10^6$ load cycles) of from 380 to $550\,N \cdot mm^{-2}$ at $t = 20\,°C$ and from 250 to $390\,N \cdot mm^{-2}$ at $t = 500\,°C$ [155].

The hardness of high speed tool steels, for example S10-4-3-10, can also be increased from about 875 HV0.3 with conventional hardening to 1100 HV0.3 by re-melting and tempering, which leaves the core of the material still in a softened condition. The primary carbides normally present in hardened high speed tool steels dissolve on re-melting and a very tough supersaturated matrix is formed [156]. Similar increases in hardness can also be achieved with both lamellar and spheroidal graphitic cast iron (Fig. 299 and 300 respectively). The small separation between the dendrites in the ledeburitic microstructure of the re-melted zone (Fig. 300, top) increases the wear resistance of this material considerably [157]. In cast irons, the surface of the metal is best re-melted by scanning the electron beam across it (Fig. 301), with the beam power and the movement of the workpiece being controlled together.

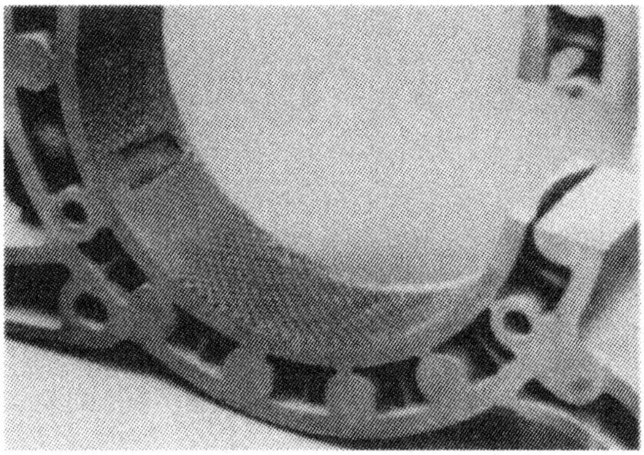

Fig. 301 Cast iron motor component showing remelted spots.

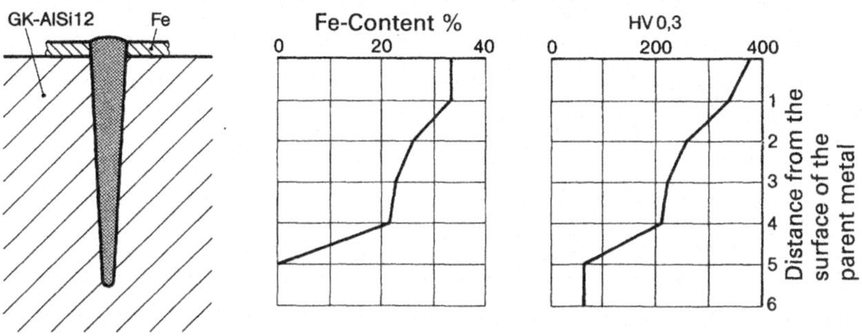

Fig. 302 Remelting of GK-AlSi 12 with iron: iron concentration and hardness traces [157].

Another interesting possibility is re-melting with addition of certain extra alloying elements. Using this technique both considerable grain refinement and increases in hardness can be achieved with aluminium alloys, e.g. AlSi 12 (Fig. 302). To do this the surface of the workpiece is covered with unalloyed steel or high alloy Cr-Ni steel, either in sheet form or electrochemically deposited, so that depending upon the volumes re-melted, the thickness of the alloying material added and the chemical composition of the respective aluminium parent material, highly metallurgically saturated fusion zones are formed.

The main requirement, however, is that the rate of cooling be as high as possible after re-melting. For this reason point re-melting is the most suitable, although linear and surface re-melting can also be carried out successfully.

13.4 Drilling, perforating

During electron beam welding the keyhole permits the beam to penetrate deeply into a material (the deep welding effect), whilst the vaporisation taking place can be used to remove material. This alone would not permit holes to be made in a material, however, since with the normal settings used for welding,

Fig. 303 Electron beam drilled gas turbine blade in high temperature nickel alloy. Wall thickness 2–3 mm, diameter of holes 0.5 mm.

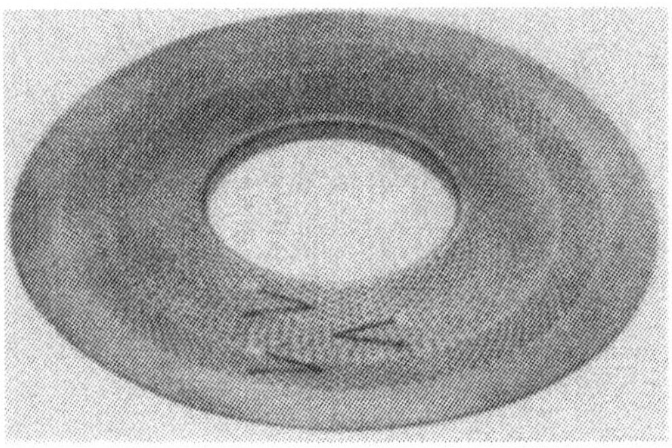

Fig. 304 Cobalt alloy disc with angled perforations. Wall thickness 1 mm, diameter of perforations 0.9 mm.

on switching off the beam the melt would close the hole formed because of surface tension. A technically useful removal of material is only achieved once:

- vaporisation is increased by increasing the power density at the focus of the beam from about 10^6 to 10^7–10^8 W · cm^{-2}; and
- the amount of molten material at any given time is kept as small as possible by pulsing the beam.

With metals other high vapour pressure materials are placed on the side of the workpiece from which the beam emerges, as an aid to blowing away the molten material produced.

Thermal removal of material with an electron beam is at present mainly used for drilling (multi-impulse technique) and perforating (single impulse technique) materials, both metals and non-metals (ceramics, plastics), which would otherwise be difficult to machine using conventional drilling processes. With this technique the hardness of the material is just as irrelevant as the stiffness of the workpiece or the angle of the axis of the required hole to the surface of the material (> 25°), since in electron beam drilling or perforating no mechanical forces are applied. In addition, being a thermal tool, the electron beam does not wear and thus always works the material with the same precision [158].

Thermal removal of material with an electron beam is at present mainly used turbine blade. The wall of the blade is perforated with a large number of holes to assist cooling, these being located in varying densities over the surface of the blade. For this type of drilling operation, the beam and the movement of the workpiece need to be controlled together simultaneously by computer [159]. Thus in electron beam perforating the workpiece remains in motion continuously. During the working phase (the pulse time) the electron beam is bent and for a short time moves with the workpiece. In the example shown in Fig. 304, the needles indicate the various directions of the perforations.

14 A comparison of electron beam and laser welding

Because of their own individual characteristics, there is no doubt that both electron beam and laser welding exceed the limits of power and efficiency of other welding processes. To make use of these advantages in manufacturing practice, the unavoidable question arises as to when it becomes technically useful and economically effective to weld with an electron beam and when using a laser?

The state of the art of process and machine development is such that these two sources of beam energy compete with one another only across a very limited range of applications. Because of the power of the lasers currently in use in manufacturing industry, a realistic comparison is only possible for welds less than 10 mm thick [160]. With weld thicknesses of $t < 3$ mm the CO_2 laser has a broad spectrum of use because of the lower costs involved when compared with electron beam welding. For welds of between $t = 3$ to 5 mm thick a careful comparison of the technical advantages and disadvantages, as well as the costs, of the two processes is required. Above $t > 5$ mm, however, welding with an electron beam becomes dominant [161]. In Table 11 a number of characteristics of the two processes are compared, which may help in choosing the more appropriate [162]. In most welding applications these general considerations will not replace the need for specific individual assessments.

Table 11 Comparison of technical characteristics of electron beam and laser welding

Term	Electron beam welding	Laser welding
Type of beam	High voltage accelerated electrons	Monochromatic coherent light
Cooling of the beam source	None	Water at ±1°C constant temperature
Wearing components	Cathode filament	Laser gas (CO_2, N_2, He), optical elements, decoupling window
Beam control	Electromagnetic lenses with variable focal length of 100–1500 mm Electromagnetic deflection system to correct the spot position and to deflect and oscillate the beam	Parabolic mirror with constant focal length in the range 100–300 mm Water cooled guidance mirror electrically or pneumatically adjusted
Welding at more than one location	By beam deflection (max. angle of deflection: 5°)	Guidance mirror to a number of remote welding stations.
Beam aperture	Variable, small	Constant, large

Table 11 (Cont.)

Term	Electron beam welding	Laser welding
External influences on the beam	Spot position can change due to magnetic fields set up by workpiece	Changes in the spot caused by bad adjustment of the mirror, effects caused by the plasma
Automatic seam tracking	By sensing the joint and measuring the reflected electrons	At present not possible
Welding environment	Evacuated working chamber	Atmosphere with additional shielding gas
Limitations on the dimensions of the workpiece	Maximum dimensions of the workpiece and length of weld determined by the dimensions of the working chamber	None; maximum distance between the beam source and point of welding ca. 10 to 15 m
Time for each welded workpiece	Welding time + handling time + evacuating time; evacuating time varies according to the size of working chamber and type of welding machine: Continuous machine = 0 Batch production = 10 to 40 s Chamber machine = 1 to 10 min	Welding time + handling time, no evacuating time
Consumables	None	Working (shielding) gas
Protective measures	Lead glass windows and lead shielding against X-rays at $U_B > 60\,kV$	Acrylic glass against dazzling
Beam penetration	Deep welding effect	Deep welding effect
Process control	Constancy of electrical and mechanical parameters, e.g. beam power, focal position, welding speed. etc. by control circuitry	Measurement of the laser beam power calorimetrically after decoupling at the resonator, and adjustment if necessary
Max. beam powers of different machines	3; 6; 10 and 15 kW, individually 30, 60 and 100 kW	1 to 10 kW, individually to 25 kW
Weld thicknesses as a function of beam power Material: Steel	P = 3 kW t = 12 mm P = 5 kW t = 20 mm P = 10 kW t = 40 mm P = 20 kW t = 80 mm P = 30 kW t = 120 mm P = 60 kW t = 240 mm	P = 3 kW t = 4 mm P = 5 kW t = 6 mm P = 10 kW t = 12 mm P = 20 kW t = 15 mm
Weld seam geometry	$t:b \leqslant 40:1$	$t:b \leqslant 10:1$

For example the economic viability of the process, the energy and material costs involved, the capital investment required, the number of components to be welded and the quality or flexibility required from the process are all aspects which require careful consideration. This must be from all points of view, and also by different specialists such as design engineers, welding engineers, marketing departments and production managers, before either of these powerful welding processes can be integrated into the manufacturing process.

References

[1] Bunshah, R. F.: The history of electron beam technology, in: Bakish, R., Introduction to electron beam technology, J. Wiley & Sons, Inc., New York-London 1962.

[2] Steigerwald, K. H.: Materialbearbeitung mit Elektronen. Physik. Verhandlungen 4 (1953), H.6, p. 123.

[3] Steigerwald, K. H.: 25 Jahre Elektronenstrahltiefschweißen. Schw. Schn. 35 (1983), H.12, p. 589/92.

[4] Maleka, A, A. H.: Electron beam welding. The Wdg. Inst. McGraw Hill Publishing Co. Ltd. 1971.

[5] Arata, Y.: Development of ultra high energy density heat sources and their application to heat processing. Trans. Jpn. Res. Inst. 13 (1984), No. 1, p. 121/45.

[6] Steigerwald, K. H.: Entwicklungen auf dem Gebiet des Elektronenstrahlschweißens. DVS-Bericht, Bd. 58, p. 13/15. DVS-Verlag, Düsseldorf 1979.

[7] Anon: Die Verfahren der Schweißtechnik. DVS-Verlag, Düsseldorf 1974.

[8] Steigerwald, K. H. Elektronenstrahlen als thermisches Werkzeug. Neue Züricher Zeitung, 27.2.1963, p. 763/66.

[9] Schiller, S., et al.: Elektronenstrahltechnologie. Wissenschaftliche Verlagsgesellschaft, Stuttgart 1977.

[10] Mayer, H. G.: Einführung in die Grundlagen des Verfahrens. Lecture at the Technischen Akademie Esslingen, 1987.

[11] Dorn, L.: Untersuchungen zum Elektronenstrahl-Schweißvorgang und zum Festigkeitsverhalten elektronenstrahlgeschweißter Metallverbindungen. Dissertation TU Hannover, 1969.

[12] Jüptner, W.: Untersuchungen zum Einbrandverhalten eines Elektronenstrahles unter Berücksichtigung der Strahlgeometrie. Dissertation TU Hannover, 1978.

[13] Mayer, H. G., and F. Noller: Physikalische und technologische Grundlagen des Elektronenstrahlschweißens. Lecture at the Technischen Akademie Esslingen, 1987.

[14] Noller, F.: Ein Elektronenstrahl hoher Richtwirkung als wesentliche Voraussetzung zum Elektronenstrahltiefschweißen. Dissertation Universität Stuttgart 1979.

[15] Noller, F., and H. G. Mayer: Elementarer Zusammenhang zwischen Strahlparametern, elektonenoptischen Kenngrößen und Nahtprofil beim Elektronenstrahlschweißen mit Triodensystemen. DVS-Bericht Bd. 44, p. 29/34. DVS-Verlag, Düsseldorf 1977.

[16] Jucker, J., et al.: Parameterstudie zum Entwurf von Elektronenstrahl-Schweißkanonen im Leistungs-bereich von 30 kW. DVS-Bericht Bd. 63, p. 31/36 DVS-Verlag, Düsseldorf 1980.

[17] Arata, Y.: Evaluation of beam characteristics by the AB test method. IIW-Doc. IV-340-83.

[18] Dumonte, P., et al.: Experimentelle Überprüfung der theoretischen Eigenschaften

des Elektronen-strahles in einer für das Schweißen verwendeten Kanone. DVS-Bericht Bd. 26, p. 15/18. DVS-Verlag, Düsseldorf 1973.

[19] Kirner, K., et al.: Optimierung der Elektronenstrahl-Justierung durch tomographische Messung der Stromdichteverteilung. DVS-Bericht Bd. 99, p. 5/10. DVS-Verlag, Düsseldorf 1985.

[20] Sickmann, J. G., and Th. M. B. Schoenmakers: New microprobe techniques for measuring the current distribution in an electron beam used for welding. IIW-Doc. IV-153-74.

[21] Depner, K.: Automatisches Messon der Leistungsdichteverteilung des Elektronenstrahles in Elektronenstrahlschweißmaschinen. DVS-Bericht Bd. 99, p. 10/14. DVS-Verlag, Düsseldorf 1985.

[22] Panten, M., et al.: Diagnosis of high energy density electron beams on electron beam welding machines with a new compact measuring equipment. 4 Int. Colloquium Welding and Melting by Electrons and Laser Beam, Cannes, Sept. 1988.

[23] Panten, M.: Beitrag zur Realisierung eines universellen Maschinendiagnosesystems und eines adaptierten Fugennachführsystems für das Elektronenstrahlschweißverfahren, Dissertation RWTH Aachen, 1988.

[24] Visser, A.: Zur Wirkungsweise des Werkstoffabtrages mit Elektronen-und Photonenstrahlen. DVS-Bericht Bd. 4, p. 173/213. DVS-Verlag, Düsseldorf 1968.

[25] Mauer, K. O.: Relationship between secondary electron current frequency and current in the work during electron beam welding and geometry of the weld. Autom. Wdg. (1982), H. 3, p. 47/49.

[26] Schiller, S.: Nutzung von Rückstreuelektronen zur Positionierung und Oberflächenabbildung beim Elektronenstrahlschweißen. DVS-Bericht Bd. 34, p. 41/46. DVS-Verlag, Düsseldorf 1975.

[27] Matting, A., and G. Sepold.: Energiebetrachtungen zum Schweißen mit Elektronenstrahlen. DVS-Bericht Bd. 4, p. 67/82. DVS-Verlag, Düsseldorf 1968.

[28] Dorn, L.: Über die Wechselwirkungen zwischen Elektronenstrahl und Materie. DVS-Bericht Bd. 5, p. 9/25. DVS-Verlag, Düsseldorf 1969.

[29] Eggers, H., and J. Ruge.: Das Energiespektrum der rückgestreuten Elektronen beim Schweißen mit dem Elektronenstrahl. DVS-Bericht Bd. 12, p. 51/59. DVS-Verlag, Düsseldorf 1971.

[30] Tong, H., and W. H. Giedt.: A dynamic interpretation of electron beam welding. Wdg. Res. Suppl. (1970), H. 6, p. 259s/66s.

[31] Schauer, D. A., and W. H. Giedt.: Prediction of electron beam welding spiking tendency. Wdg. Res. Suppl. (1978), H. 7, p. 189s/95s.

[32] Eggers, H., and J. Ruge.: Ausbildung der Dampfkapillare beim Elektronenstrahlschweißen. Schw. Schn. 24 (1972), H. 2, p. 200/04.

[33] Arata, Y.: Some dynamic aspects of weld molten metal in electron beam welding. Trans. Jpn. Wdg. Res. Inst. 2 (1973), H. 2. p. 22/32.

[34] Knipp. H. G., and J. Ruge.: Untersuchungen zur Kapillarendynamik und Nahtausbildung beim Elektronenstrahlschweißen. DVS-Bericht Bd. 34, p. 30/34. DVS-Verlag, Düsseldorf 1975.

[35] Hesse, P.: Die Schmelzbadbewegung beim Elektronenstrahlschweißen – Stand der Erkenntnis und Untersuchungsergebnisse. DVS-Bericht Bd. 44, p. 9/16. DVS-Verlag, Düsseldorf 1977.

[36] Schebasta, W.: Untersuchungen zur Kapillarendynamik beim Elektronenstrahlschweißen. DVS-Bericht Bd. 44, p. 17/21. DVS-Verlag, Düsseldorf 1977.

[37] Hehl, D., et al.: Beitrag zur Theorie der Dampfkapillare beim Elektronenstrahlschewißen. DVS-Bericht Bd. 44, p. 21/28. DVS-Verlag, Düsseldorf 1977.

[38] Arata, Y.: Tandem electron beam welding. Trans. Jpn. Wdg. Res. Inst. 11 (1982), H. 1. p. 1/5.
[39] Tsukamoto, S., et al.: Effect of beam current on humping bead formation in electron beam welding. Trans. Nat. Res. Inst. Met. 26 (1984), H. 2. p. 35/50.
[40] Eichhorn, F., and D. Schumacher.: Beitrag zum Aufschmelz- und Erstarrungsprozeß beim Elektronenstrahlschweißen. DVS-Bericht Bd. 26, p. 27/33. DVS-Verlag, Düsseldorf 1973.
[41] Tsukamoto, S., and H. Irie.: Welding defects and molten metal behaviour in low speed electron beam welding. Wdg. in the World 23 (1985), No. 5/6, p. 130/40.
[42] Anderl, P., and W. Scheffels.: Controlled EB welding of heavy sections with a double lens EB gun. 3. Internat. Coll. Welding and Melting by Electrons and Laser Beams, Lyon, Sept. 1983.
[43] Arata. Y.: Fundamental features of advanced laser and electron beam technology. 4. Internat. Coll. Welding and Melting by Electrons and Laser Beams, Cannes, Sept. 1988.
[44] Dietz, et al.: Berechnung der Einschmelztiefe beim Elektronenstrahlschweißen. VDI-Z. 112 (1970), No. 17, p. 1181/86.
[45] Wei, P. S., and W. H. Giedt.: Surface tension gradient driven flow around an electron beam welding cavity. Wdg. Res. Suppl. (1985), H. 9, p. 251-s/59-s.
[46] Adam, P.: Einfluß der Absenkcharakteristik auf die Qualität der Elektronenstrahlschweißungen im Auslauf. DVS-Bericht Bd. 34, p. 71/73. DVS-Verlag, Düsseldorf 1975.
[47] Schumacher, D., et al.: Strahldiagnostische Untersuchungen beim Schweißen mit hoher Strahlleistung im Hoch-und Feinvakuum. DVS-Bericht Bd. 34, p. 14/21. DVS-Verlag, Düsseldorf 1975.
[48] Kita, H., et al.: Application of high power electron beam welding to heavy structures. Doc. IIS/IIW-681-81, Wdg. in the World 19 (1981), No. 516, 79/91.
[49] Cazes, R., et al.: Contrôle de la décharge électrique (flash) dans les canons de soudage par faisceau électronique. DVS-Bericht Bd. 99, p. 55/58. DVS-Verlag, Düsseldorf 1985.
[50] Panzer, S., et al.: Vermeidung technologisch- und werkstoffbedingter Hochspannungsüberschläge in der Elektronenkanone beim Elektronenstrahlschweißen. ZIS-Mitt. 23 (1981), H. 1, 34/38.
[51] Göhre, H. R.: Hochspannungsüberschläge während der Bearbeitung von Aluminiumkolben mit dem Elektronenstrahl. ZIS-Mitt. 26 (1984), H. 1, 69/72.
[52] Passoja, D. E.: Penetration of solids by high power density electron beams. Brit. Wdg. J. (1967), H. 1, p. 13/16.
[53] Koch, F., and G. Neubert.: Einbrandtiefe beim Elektronenstrahlschweißen mit großen Leistungen. ZIS-Mitt. 9 (1967), H. 2, 1625/33.
[54] Eggers, H., and J. Ruge.: Selbsttätiges Fokussieren beim Elektronenstrahlschweißen mit großen Leistungen. Schw. Schn. 23 (1971), H. 2, p. 45/47.
[55] Ruge, J., and H. Eggers.: Werkstoffauswurf beim Elektronenstrahlschweißen. DVS-Bericht Bd. 21, p. 15/21. DVS-Verlag, Düsseldorf 1971.
[56] Eggers, H., et al.: Einfluß von Strahlleistung, Schweißgeschwindigkeit und Strahloszillation auf die selbsttätige Fokussierung beim Elektronenstrahlschweißen. DVS-Bericht Bd. 26, p. 19/21. DVS-Verlag, Düsseldorf 1973.
[57] Mayer, R., and F. Noller.: Automatische Fokussierung des Elektronenstrahles beim Elektronenstrahlschweißen. DVS-Bericht Bd. 63, p. 43/47. DVS-Verlag, Düsseldorf 1980.
[58] Depner, K., et al.: Zusammenhang zwischen Strahlgeometrie und Einbrandform

bei größeren Nahtdicken an Blechen aus niedriglegiertem Stahl. DVS-Bericht Bd. 63, p. 12/19. DVS-Verlag, Düsseldorf 1980.

[59] Noller, F., and H. G. Mayer.: Elementarer Zusammenhang zwischen Strahlparametern, elektronen-optischen Kenngrößen und Nahtprofil beim Elektonenstrahlschweißen mit Triodensystemen. DVS-Bericht Bd. 44, p. 29/34. DVS-Verlag, Düsseldorf 1977.

[60] Farrel, W. J., and J. D. Ferrario.: A computer controlled wide bandwidth deflection system for electron beam welding and heat treating. Wdg. J. (1987), No. 10, p. 41/49.

[61] Sievers, E. R., and H. D. Steffens.: Das Fügen dicker Querschnitte in der Tulpennahttechnik. DVS-Bericht Bd. 99, p. 151/55. DVS-Verlag, Düsseldorf 1985.

[62] Mayer, R., et al.: Zur analogen Strahlführung beim Elektronenstrahlschweißen. DVS-Bericht Bd. 12, p. 19/24. DVS-Verlag, Düsseldorf 1971.

[63] Mayer, H. G., et al.: Anwendung der Elektronenstrahl-Ablenktechnik beim Fügen von Bauteilen und zur gezielten Wärmeeinbringung beim Schweißen größerer Wanddicken. DVS-Bericht Bd. 34, p. 55/60. DVS-Verlag, Düsseldorf 1975.

[64] Wiesner, P., and H. Ehrhardt.: Elektronenstrahlschweißen mit Strahlpendeln. ZIS-Mitt. 25 (1983), H. 1, p. 17/28.

[65] Depner, K., et al.: Elektronenstrahlschweißen dickwandiger Stahlbleche für den Behälter- und Apparatebau. DVS-Bericht Bd. 99, p. 45/50. DVS-Verlag, Düsseldorf 1985.

[66] Anderl, P.: Elektronenstrahl-Tiefschweißen mit dem 60-kW-Hochleistungsgenerator. trennen + fügen (1985), H. 14, p. 38/42.

[67] Anderl, P., et al.: Fortschritte beim Elektronenstrahlschweißen sehr tiefer Nähte in verschiedenen Fugenlagen. DVS-Bericht Bd. 99, p. 50/54. DVS-Verlag, Düsseldorf 1985.

[68] Dietrich, W., et al.: Aktueller Stand der Elektronenstrahl-Schweißtechnologie in der Produktion. DVS-Bericht Bd. 99, p. 67/73. DVS-Verlag, Düsseldorf 1985.

[69] Schauer, D. A.: Electron beam welding cavity temperature distribution in pure metals and alloys. Wdg. Res. Suppl. (1978), H. 5, p. 127-s/33-s.

[70] Steigerwald, K. H., and D. König.: Neue Ergebnisse auf den Gebieten der Werkstoffbearbeitung und-behandlung mit Elektronenstrahlen. DVS-Bericht Bd. 21, p. 9/14. DVS-Verlag, Düsseldorf 1971.

[71] Dorn, L.: Zur Temperaturverteilung beim Bohren und Schweißen mit Elektronenstrahlen. DVS-Bericht Bd. 12, p. 65/69. DVS-Verlag, Düsseldorf 1971.

[72] Kreye, H., et al.: Zum Entstehen von Erstarrungsringen in Schweißnähten. Schw. Schn. 27 (1975), H. 7, 253/57.

[73] Dorn, L.: Untersuchungen zum Korrosionsverhalten des austenitischen CrNi-Stahles × 12 CrNi 18 8. DVS-Bericht Bd. 4, p. 95/113. DVS-Verlag, Düsseldorf 1968.

[74] Merkblatt DVS 1002: Schweißeigenspannungen, Teil 1: Einteilung, Benennung, Erklärungen (1983), Teil 2: Verfahren zur Verringerung von Schweißeigenspannungen (1986).

[75] Poje, R.: Schweißeigenspannungen in Elektronenstrahl- und Unterpulver-Schweißverbindungen von Stählen bei verschiedenen Blechdicken. Dissertation RWTH Aachen, 1984.

[76] Matsuda, F., and K. Nakata.: A new test specimen for self-restraint solidification crack susceptibility testing of electron beam welding beads. Trans. Jpn. Wdg. Res. Inst. 11 (1982), H. 2, p. 87/94.

[77] Russell, J. D.: The development of an electron beam weld solidification cracking test. Wdg. Res. Int. 9 (1979), H. 6, p. 1/21.
[78] VdTÜV-Merkblatt 1159 Verfahrensprüfung und Arbeitsprüfung für elektronenstrahlgeschweißte Bauteile.
[79] Depner, K., et al.: Investigations in electron beam welding of thick walled heat resistant steels. 4 Int. Coll. Welding and Melting by Electron and Laser Beam, Cannes, Sept. 1988.
[80] Thier, H.: Delta-Ferrit und Heißrisse beim Schweißen chemisch beständiger Stähle. DVS-Bericht Bd. 41, p. 100/04. DVS-Verlag, Düsseldorf 1976.
[81] Lippold, J. C.: Centerline cracking in deep penetration electron beam welds in type 304L stainless steel. Wdg. Res. Suppl. 64 (1985), H. 5, p. 127-s/36-s.
[82] Tauschitz, C., et al.: Untersuchungen zur Elektronenstrahl-Schweißbarkeit einer hochwarmfesten Eisenbasis-Superlegierung (A286). Z. Metallkd. 73 (1982), H. 11, p. 717/24.
[83] Meyer, W. E., and H. Schultz.: Festigkeitseigenschaften elektronenstrahlgeschweißter hochwarmfester Legierungen. Schw. Schn. 19 (1967), H. 3, p. 111/16.
[84] Schultz, H.: Festigkeitsverhalten einiger elektronenstrahlgeschweißter hochfester Baustähle. Schw. Schn. 19 (1967), H. 3, p. 591/98.
[85] Hiller, W.: Schweißen und thermisches Behandeln von Gußwerkstoffen mit dem Elektronenstrahl. Gießerei 63 (1976), H. 11, p. 316/21.
[86] Sievers, E, -R., et al.: Elektronenstrahlschweißen von Kupfer. DVS-Sericht Bd. 99, p. 115/19. DVS-Verlag, Düsseldorf 1985.
[87] Sievers, E. -R.: Personal communication.
[88] Iversen, K. V., et al.: Herstellung von Limitern zur Plasmabegrenzung im Tokamak von JET. DVS-Bericht Bd. 75. DVS-Verlag, Düsseldorf 1982.
[89] Morochko, V. P., et al.: A classification of high-temperature nickel alloys by their electron beam weldability. Autom. Wdg. 33 (1980), H. 12, p. 29/31.
[90] David, S. A., et al.: Welding and weldability of nickel-iron aluminides. Wdg. Res. Suppl. 64 (1985), H. 1, p. 22-s/28-s.
[91] Anderl, P., et al.: Elektronenstrahlschweißen von Aluminiumwerkstoffen. Aluminium 62 (1986), H. 7, p. 596/600.
[92] Anon: Results of the co-operative program on electron beam welding of aluminium alloy AlZnMg1. Wdg. in the World 22 (1984), p. 186/94.
[93] Bondarev, A. A.: Zur Schweißbarkeit von Aluminiumgußlegierungen und zur Technologie des Elektronenstrahlschweißens von Kolben mit Kühlraum. ZIS-Mitt. 27 (1985), H. 1, p. 50/71.
[94] Ruge. J, and P. Lutze.: Eignung verschiedener Schweißverfahren zum Schweißen von Aluminium-Druckguß des Legierungstyps Aluminium-Silicium Gießerei 72 (1985), H. 12, p. 360/366.
[95] Anon: Aluminium-Taschenbuch, 14. Auflage, Aluminium-Verlag, Düsseldorf 1983.
[96] Zaderii, B. A., et al.: Formation of electron beam welds in 70-260 mm thick slabs of titanium. Aut. Wdg. (1985), H. 7, p. 50/53.
[97] Merkblatt DVS 2713.: Schweißen von Titanwerkstoffen (1987).
[98] Schultz, H.: Schweißen von Sondermetallen. Fachbuchreihe Schweißtechnik. Bd. 59. DVS-Verlag, Düsseldorf 1971.
[99] Lison, R.: Einsatzmöglichkeiten thermischer Fügeverfahren für Sondermetalle. Radex-Rundschau (1983), H. 1/2, p. 99/117.
[100] Lison, R.: Energiereich und schnell. Maschinenmarkt 92 (1986), H. 49, p. 37/40.
[101] Lamatsch, H., et al.: Über das vakuumdichte Schweißen von Molybdän. DVS-Bericht Bd. 12, p. 25/28. DVS-Verlag, Düsseldorf 1971.

[102] Lison, R.: Schweißen von Molybdän und seinen Legierungen. Schw. Schn. 38, (1986), H. 9, p. 437/41.

[103] Hiller, W.: Metallurgische Gesichtspunkte bei der Anwendung des Elektronenstrahlschweißens. Lecture at the Technischen Akademie Esslingen, 1984.

[104] Wiesner, P., and H. Hoppe.: Vorausbestimmung der Schweißbarkeit mit dem Mikrorechner. ZIS-Mitt. 27 (1985), H. 1, p. 29/32.

[105] Scheibe, W.: Das Verschweißen unterschiedlicher Werkstoffe, bei denen nach dem Zustandsdiagramm intermetallische Phasen auftreten. DVS-Bericht Bd. 4, p. 17/43. DVS-Verlag, Düsseldorf 1968.

[106] Scheibe, W.: Die Beeinflussung intermetallischer Phasen beim Elektronenstrahlschweißen von dickeren Blechen und runden Querschnitten aus verschiedenartigen Werkstoffen sowie deren Einsatz bei höheren Temperature. DVS-Bericht Bd. 5, p. 51/56. DVS-Verlag, Düsseldorf 1969.

[107] Metzger, G., and R. Lison.: Schweißverbindungen an unterschiedlichen Werkstoffen. Bericht KFA Jülich, July-1011-RW, Oct. 1973.

[108] Ruge, J., and M. Witt.: Einfluß von Strahlführung, Drahtzufuhr und Nahtgeometrie auf die Werkstoffvermischung beim Elektronenstrahlschweißen mit Schweißzusatz. DVS-Bericht Bd. 63, p. 89/94. DVS-Verlag, Düsseldorf 1980.

[109] Ruge, J., et al.: Welding of dissimilar metals by using the method of electron beam welding with filler wire. Internat. Conference Tokyo, July 1986.

[110] Lehrheuer, W.: Schweißen dicker Stahlplatten mit einem Elektronenstrahl hoher Leistung – eine Methode zum Vermeiden der durch Restmagnetismus hervorgerufenen Strahlablenkung. Schw. Schn. 37 (195), H. 2, p. 84.

[111] Akutsu, Y., et al.: Application of electron beam welding to steam turbine diaphragms. DVS-Bericht Bd. 63, p. 63/69. DVS-Verlag, Düsseldorf 1980.

[112] Blakeley, P. J., et al.: An investigation into the origin and effect of magnetic fields in EB welding. Report by The Welding Institute Abington No. 172/1982.

[113] Okamura, T. S., et al.: Experiments on ultrasonic in-process monitoring of weld defects during electron beam welding. Int. Conference Power Beam Technology, Brighton 1986. The Welding Institute Abington p. 431/37.

[114] Merkblatt DVS 3201: Grundsätze für das Konstruieren elektronenstrahlgeschweißter Bauteile (1986).

[115] DIN 65 118 Part 2, Draft Oct. 1988.: Luft- und Raumfahrt; Geschweißte metallische Bauteile, Konstruktionshinweise.

[116] Beckert, M., and G. Neuberg.: Elektronenstrahlschweißen mit Zusatzwerkstoff. DVS-Bericht Bd. 12, p. 39/42. DVS-Verlag, Düsseldorf 1971.

[117] Grimm, H. W., and J. Ruge.: Untersuchungen zur Technologie und Metallurgie des Elektronenstrahlschweißens mit Zusatzwerkstoff. DVS-Bericht Bd. 34, p. 35/40. DVS-Verlag, Düsseldorf 1975.

[118] Yasunda, K., et al.: Wire feeders boost EB welding's potential. Wdg. and Met. Fabrication, April 194, p. 101/103.

[119] Anderl, P., et al.: Automatische Steuerung des Elektronenstrahl-Schweißprozesses mit Schweißzusatz in Abhängigkeit von der Werkstückabmessung und magnetischen Störfeldern. DVS-Bericht Bd. 63, p. 95/102. DVS-Verlag, Düsseldorf 1980.

[120] Oestmann, C., et al.: Elektronenstrahl-Engspaltschweißen von Blechen bis 100 mm Dicke mit Elektronenstrahlen geringer Leistung unter Anwendung von Schweißzusatz und Mehrlagentechnik. DVS-Bericht Bd. 99, p. 134/37. DVS-Verlag, Düsseldorf 1975.

[121] Witt, M.: Untersuchungen zum Elektronenstrahlschweißen mit Zusatzwerkstoff. Schw. Schn. 38 (1986), H. 7, p. 337/39.

[122] Grimm, H., and J. Ruge.: Elektronenstrahlschweißen mit drahtförmigem Zusatz-

Einsatz an Stahl mit Blechdicken bis 30 mm. DVS-Bericht Bd. 44, p. 39/43. DVS-Verlag, Düsseldorf 1977.
[123] Fletcher, M. D.: Distortion in electron beam welding. Wdg. and Met. Fabrication, Aug. 1971.
[124] Wabersich, E.: Längs- und Querschrumpfungen an elektronenstrahlgeschweißten Titanblechen. DVS-Bericht Bd. 5, p. 59/62. DVS-Verlag, Düsseldorf 1969.
[125] Decker, I., et al.: Einfluß der Nahtgestaltung auf die Fertigungsgenauigkeit beim Laserstrahlschweißen im Vergleich zum Elektronenstrahlschweißen. DVS-Bericht Bd. 109, p. 134/37. DVS-Verlag, Düsseldorf 1987.
[126] Naderi, H.: Bleibende und vorübergehende Verformungen in elektronenstrahlgeschweißten Teilen. ZIS-Mitt. 1/79, p. 57/64.
[127] Wiesner, P., et al.: Elektronenstrahlschweißen von Getrieberädern. ZIS-Mitt. 1/80, p. 18/27.
[128] Hardenberg, K.: Verformungen beim Elektronenstrahlschweißen. ZIS-Mitt. 1/75, p. 43/50.
[129] Kirpach, I. P., et al.: Deformation of gas turbine rotor discs during electron beam welding. Automatic Wdg. 12/1984, p. 31/34.
[130] Hardenberg, K.: Untersuchungen zum Elektronenstrahlschweißen axialer Rundnähte. Schweißtechn. (Berlin) 30 (1980), H. 7, p. 322.
[131] Drews, P., and B. Spies.: Prozeßregelung der Durchschweißung mit dem Elektronenstrahl. DVS-Bericht Bd. 44, p. 64/70. DVS-Verlag, Düsseldorf 1977.
[132] Eichhorn, F., et al.: Durchschweißregelung für das Elektronenstrahlschweißen. Industrie-Anzeiger 104 (1982), No. 69, p. 30/32.
[133] Mauer, K. O.: Systeme zur Strahl-Stoß-Positionierung beim Elektronenstrahlschweißen. Schweißtechn. Berlin 32 (1982), H. 8, p. 368/73.
[134] Carrol, M. J., and D. E. Powers.: Automatic joint tracking for CNC programmed electron beam welding. Wdg. J. 64 (1985), H. 8, p. 34/38.
[135] Koch, D., et al. Automatische Strahlnachführung – ein weiterer Schritt auf dem Wege zur Automatisierung der Arbeit an der Elektronenstrahl-Schweißmaschine. DVS-Bericht Bd. 26, p. 23/26. DVS-Verlag, Düsseldorf 1973.
[136] Entwurf Merkblatt DVS 3202: Elektronenstrahl- und Maschinensteuerungen.
[137] Müller, A., et al.: Elektronenstrahlschweißen, computernumerisch gesteuert. Schw. Schn. 39 (1987), H. 3, p. 111/16.
[138] Kita, H., et al.: Technological and economical evaluation of EB welding heavy components of electrical machines. Japan-German Colloquium, Welding and Economics, Tokyo (Jp), Session 3 (1983), p. 147/54.
[139] Lehrheuer, W.: Entwicklungsstand des Elektronenstrahlschweißens großer Wanddicken. DVS-Bericht Bd. 63, p. 48/52. DVS-Verlag, Düsseldorf 1980.
[140] Anderl, P., et al.: Elektronenstrahlschweißen von Großwerkstücken mit Mobilem-Vakuum-Aggregat unter praxisnahen Bedingungen. DVS-Bericht Bd. 63, p. 53/59. DVS-Verlag, Düsseldorf 1980.
[141] Arata, Y., et al.: Development of electron beam welding in a local vacuum chamber system. Trans. Jpn. Wdg. Res. Inst. 11 (1982), No. 1, p. 25/29.
[142] Anon: Information from the Vereinigte Edelstahlwerke.
[143] Schütz, W., and W. Oberparleiter.: Influence of welding flaws on the fatigue strength of electron beam weldments in TiAl6V4. AGARD Conference Proceedings 398, Oberammergau 11. – 13. Sept. 195.
[144] Merkblatt DVS 2401: Bruchmechanische Bewertung von Fehlern in Schweißverbindungen; Teil 1: Grundlagen und Vorgehensweise (1982), Teil 2: Praktische

Anwendung (1989). Fachbuchreihe Schweißtechnik Bd. 101. DVS-Verlag, Düsseldorf 1989.
[145] Stegherr, R.: Flugtriebwerksgetriebe unter Anwendung moderner Fertigungsverfahren. VDI-Report No. 2332 (1979), p. 113/18.
[146] Reinecken, K.: Darstellung von thermischen Verfahren am Beispiel der Triebwerksteile-Reparatur. Mitt. Deutsche Lufthansa AG, Hamburg.
[147] Höll, H.: Elektronenstrahlschweißen von Getrieberädern in der Serienfertigung. DVS-Bericht Bd. 63, p. 132/36. DVS-Verlag, Düsseldorf 1980.
[148] Kluth, K. H., and E. Meyer.: Elektronenstrahlschweißen im Kraftfahrzeugbau – dargestellt am Beispiel der Lkw-Hinterachse. DVS-Bericht Bd. 63, p. 129/32. DVS-Verlag, Düsseldorf 1980.
[149] Herden, G., et al.: Fertigungskosten beim Elektronenstrahlschweißen. ZIS-Mitt. 17 (1975), H. 1, p. 22/34.
[150] Merkblatt DVS 2706: Betriebliche Ausbildung von Bedienpersonal für Elektronenstrahl-Schweißmaschinen.
[151] Entwurf Richtlinie DVS 1187: DVS-Lehrgang Fachkraft für die Metallbearbeitung durch Laserstrahl.
[152] Schultz, H.: Technisches Regelwerk und betriebliche Voraussetzungen für das Elektronenstrahl schweißen. DVS-Bericht Bd. 99, p. 124/27. DVS-Verlag, Düsseldorf 1985.
[153] Steigerwald, K. H., and D. König.: Neue Ergebnisse auf den Gebieten der Werkstoffbearbeitung und -behandlung mit Elektronenstrahlen. DVS-Bericht Bd. 21, p. 9/15. DVS-Verlag, Düsseldorf 1971.
[154] Schiller, S., and S. Panzer.: Thermische Oberflächenmodifikation metallischer Bauteile mit Elektronenstrahlen. Metall 39 (1985), No. 3, p. 222/32.
[155] Hiller, W.: Steigerung der Dauerfestigkeit von Werkstücken durch örtliches Umschmelzen mit dem Elektronenstrahl. DVS-Bericht Bd. 26, p. 61/63. DVS-Verlag, Düsseldorf 1973.
[156] Hiller, W.: Aufschmelzbehandlungen metallischer Werkstoffe mit dem Elektronenstrahl – Grundlagen und Anwendung. Special publication 15/81 Messer Griesheim.
[157] Hiller, W.: Neue Wege der Oberflächenveredelung metallischer Werkstoffe durch Elektronenstrahl-Aufschmelzbehandeln. Metall 31 (1977), No. 1, p. 2/4.
[158] Steigerwald, K. H., et al.: Produktmäßige Perforation von Triebwerksteilen mittels numerisch gesteuerter Elektronenstrahlmaschinen. DVS-Bericht Bd. 44, p. 74/77. DVS-Verlag, Düsseldorf 1977.
[159] v. Dobeneck, D.: Materialbearbeitung mit Elektronenstrahlen – der Prozeß und seine industrielle Anwendung. TZ für prakt. Metallverarbeitung 69 (1975), No. 5.
[160] Anon: Laser- und Elektronenstrahl – beide sind vielversprechend. Der Praktiker 40 (1988), H. 6, p. 298.
[161] v. Dobeneck, D.: Laser- oder Elektronenstrahl? Lecture, Munich 1986.
[162] Anderl, P.: Laser- und Elektronenstrahlschweißen. DVS-Bericht Bd. 115, p. 17/26. DVS-Verlag, Düsseldorf 1988.

Photographic acknowledgements

Fig. No.	Source of Photograph
27, 28, 29	Institut für Schweißtechnische Fertigungsverfahren, RWTH Aachen
33, 35, 41, 46, 72, 73, 79, 81, 82, 93, 95, 108, 174, 186, 187, 188, 189, 190, 191, 192, 193, 195, 198, 199, 200, 202, 204, 205, 206, 207, 220, 221, 223, 227, 228, 231, 232, 252, 254, 263, 264, 267, 276, 283, 299, 300	Messer Griesheim Steigerwald Strahltechnik Gmbh, Puchheim
66, 67	Lehrstuhl für Werkstofftechnologie, Universität Dortmund
143	Institut für Schweißtechnik und Werkstofftechnologie, Technische Universität Braunschweig
209, 210, 211, 213, 214, 216, 217	Balzers Hochvakuum GmbH, Balzers
222, 245, 247, 268, 269, 270, 271	Sulzer AG, Winterthur
226, 256, 257	Daimler Benz AG, Stuttgart
241, 242, 243, 244	MTU GmbH, München
249, 250, 251	Deutsche Lufthansa AG, Hamburg
253, 284	Leybold AG, Hanau
255, 260, 261	Volkswagen AG, Kassel
273, 274	Elektronenstrahltechnik Nord, Braunschweig
288, 289, 290, 291, 292, 294, 295, 296	Beuth Verlag, Berlin/Köln

Index

CPSIA information can be obtained at www.ICGtesting.com
Printed in the USA
BVOW06*0805290416

445870BV00006B/33/P

9 781855 730502